Max von Laue

HISTORY OF PHYSICS

Translated by Ralph Oesper

Edited by Vesselin Petkov

MINKOWSKI
Institute Press

Max von Laue (9 October 1879 – 24 April 1960)

Cover photos: `https://upload.wikimedia.org/wikipedia/commons/9/97/Max_von_Laue_sign.jpg`

Published 2023

ISBN: 978-1-998902-01-9 (softcover)
ISBN: 978-1-998902-02-6 (ebook)

Minkowski Institute Press
Montreal, Quebec, Canada
http://minkowskiinstitute.org/mip/

For information on all Minkowski Institute Press
publications visit our website at
http://minkowskiinstitute.org/mip/books/

EDITORS'S PREFACE

This is a new publication of Max von Laue's famous book *History of Physics*.[1]

This is a valuable book not only because of its very informative content but particularly because it was written by a Nobel Laureate for Physics (1914). Although it covers the history of physics up to 1940, this does not make it outdated because no period of history, especially of the history of physics, can be outdated and because 1940 ("taken as the terminal date" as von Laue put it) is well after the completion of the two greatest revolutions not only in physics but in all the science – the theory of relativity and quantum mechanics.

Also, wherever appropriate, a number of notes (as footnotes or notes at the end of chapters) are added in the text to reflect recent advances and to provide additional clarification that may help to avoid confusions and misconceptions.

What makes this book even more valuable is von Laue's brilliant stile of writing. Here two examples to demonstrate that reading this book will undoubtedly be both helpful and enjoyable:

> The question is often raised as to the objectivity, the truth of scientific knowledge... Yet, it [physics] itself furnishes proof of its objective truth, proof that has overwhelming power of conviction... Whoever has been privileged to live through such an extremely surprising event, even at a consider-

---

[1]M. von Laue, *History of Physics*, translated by Ralph Oesper (Academic Press, New York 1950).

able distance, or, at least, to survey it after it has occurred, can no longer retain any doubt that the confluent theories certainly contain, if not complete truth, a substantial core of objective truth that is devoid of human embellishment. [p. 8]

Certainly, nothing in its previous history had contributed so powerfully to the establishment of respect for the young science of physics as did the Newtonian computation of the orbits of the planets. From then on, this science has been a great mental empire, which no other power may ignore with impunity. [p. 35]

The text was typeset in LaTeX by Svetla Petkova and noticed typos were corrected.

28 January 2023

Vesselin Petkov
Minkowski Institute
Montreal

PREFACE

The term "history" generally refers to political history, and hence its principal concerns are the actions and sufferings of peoples, the ups and downs of their national existence. However, there is another concept of history, at least for a minority of people. To them, the usual interpretation of history merely provides a frame for something more important, namely, the history of the mental development of humanity. Everything connoted by the almost undefinable term, *Weltanschauung,* belongs in this category. Among other things, the history of science is also included.

However, even this province of learning is itself now so extensive that no individual, not even a modern Leibnitz, would attempt to set it forth as an entirety. Consequently, it has long been customary to treat the history of medicine, astronomy, chemistry, etc., as single topics. The same holds true for the history of physics; however, no such volume has appeared in German for a good many years. What has been added to physics since 1900 – and this increment is not small – apparently has not been presented in connected fashion anywhere else either. In addition, older accounts of the early history of physics, which can be found in a number of books that were excellent in their time, were not written from a modern point of view. It was for such reasons that I yielded to the urging of good friends and decided to write a history of physics that would fit within the limits of a series of histories of the various sciences being published by the University Press at Bonn under the editorial guidance of Professor Rothacker. I decided to bring this recital down to the approximate present, i.e., in general, 1940 was taken as the terminal date.

I am well aware of the risk thus taken. It is more than likely that the future will evaluate some of this material quite differently, but I shall be content if later historians of physics even consider my way of looking at events. They can then form their own conclusions. I only hope that they do not deny that I have used objectivity and care.

The first draft of this book was written in the summer of 1943 and is accordingly a war product. Precisely because of this circumstance, the thought of the culture that is the common property of all nations, and which was then being so despicably mistreated, was ever present throughout the period of composition. It is my hope that this translation may contribute to an increasing awareness of a world-wide unity of interest, and to an intellectual union.

Göttingen, December 1949   MAX VON LAUE

CONTENTS

INTRODUCTION

History can be written from quite divergent viewpoints but still with complete adherence to the truth. There is justification for every viewpoint from which the historian can extract something of historical interest. The history of a science can likewise be treated from a variety of viewpoints. The basis of the present text is the genesis and the changes experienced by certain ideas and information that are of importance to the physics of today. Just as any political history must close before it can include the political events of the present moment, the history of a science likewise cannot deal finally with those problems which cannot be considered as definitely solved.

The extreme past can contribute very little to this report, and its accomplishments in physics can be adequately summed up in a few sentences. The Sumerians, Babylonians, and Egyptians admittedly had considerable acquaintance with single physical topics which, of course, gave the impression of being accidental, unsystematic, and not really thought through. The Greco-Roman period gave rise, among the fields of knowledge that are dealt with in this book, only to statistics, which is a branch of mechanics. Certain statements of Plato (427-347 B.C.) that have come down to us, reveal a thorough contempt for all empirical research, joined to a vigorous disparagement of efforts to remove the exalted science of mathematics from the realm of pure thinking and to desecrate it by applications to matters of actual experience. It is fully in accord with such thinking that his pupil Aristotle (384-322 B.C.) saw fit to include, in his otherwise grandiose system of natural science, only a few concepts, taken rather

1

noncritically from superficial observations, and their logical or oftentimes merely sophistical analysis. Thus even a genius like Archimedes (287-212 B.C.) remained without enduring influence. Nothing in either antiquity nor the medieval period points to any systematic scientific investigation.

The first signs of a new spirit of inquiry were given by the great voyages of discovery at the close of the fifteenth century, especially that of Christopher Columbus (1446?-1506), which culminated in his discovery of America in 1492. This brave adventurer not only knew that the earth is a sphere, a fact known even to Eratosthenes (276-195 B.C.), but he was the first to have so much faith in this idea that he made it the entire basis of his undertaking, a venture which many of his contemporaries considered foolhardy. But even in the sixteenth century which, through translations and commentaries, had successfully adopted the scientific notions of the ancients, the superior feat of Copernicus (1473-1548) received the attention of only a few, some agreeing with, others denying his heliocentric theory. It was not until the early years of the seventeenth century, when the circle of those interested in natural science became large enough, that any discussion of a continuously advancing research is really warranted. The interest in science was greatly vitalized by the then generally current effort to abandon speculative methods and tradition and to base science instead on observation, or even more, to institute carefully planned experiments. This completely new approach was regarded by many at the time as an abrupt break with the past, an idea that still persists. Was this really the case? If, in antique culture, the dominant principle was the subordination of the individual in the general scheme of things, as was proclaimed by the Greek dramatists and as was carried out by the mathematicians in their science, the new disposition toward the natural sciences was merely the logical extension of this philosophy to a field which the ancients had barely entered. Suddenly, about 1600, two new fundamental means of observation were available: the microscope and telescope. Their actual inventors are not known.

Galileo Galilei, who, unlike Copernicus, did not write solely for the scholars ("mathematicians" as he called them) but for everyone, attracted numerous pupils and followers. It was not until this time that the Copernican system came to be generally known, and the smoldering controversy about accepting it was fanned into a fierce flame. It was at least in the background when Giordano Bruno was sent to the stake in 1600, because the doctrine of the infinite extension of space and the multiplicity of worlds, which was among his alleged heresies, was a pertinent extension of the Copernican system. However, neither this execution nor the ecclesiastical interdict, which the Inquisition laid on Galileo and the Copernicans as a class, proved to have any lasting effect. The ban was finally and completely lifted at the beginning of the nineteenth century.

The eighteenth and nineteenth centuries witnessed no further attempts by the ecclesiastical and governmental authorities to interfere with the scientists; the guillotining of Lavoisier in 1794 had no connection with his scientific beliefs. This attitude of noninterference was maintained until the Hitler regime came to power; the relativity theory, in particular, was proscribed by the Nazis but this ban was lifted eventually. In general, physics was permitted to develop peacefully according to its own laws.[2] As a result, the science grew into a movement not divorced from daily life, but instead, through its technical applications, exerted a direct influence on both individuals and nations. In fact, its concepts, in a quiet but nonetheless effective manner, had such potent repercussions that even political history cannot be understood without taking these influences into account. One of the aims of this book will be to demonstrate the marked extent to which the mental structure of the man of today reflects the mental labors of the physicists of the past three or three and a half centuries.

Though the churches, in general, abstained from inter-

---

[2]Obviously, the personal lives of the physicists were intertwined with the events of their times, but this phase of the history of physics must be treated elsewhere.

fering officially, the scientific activities of the physicists have always been influenced by their private religious views. The latter, of course, were not necessarily identical with the ecclesiastical doctrines, but the philosophical attitudes of the scientists were affected, at least to some extent, by the prevailing religious thought. Kepler, Descartes, Leibniz, and Newton freely acknowledged this influence; it played a part in the principle of least action in the eighteenth century. After this period, in which Kant's philosophy proclaimed the complete independence of scientific understanding and religious belief, not much more about it is found in physical writings. However, this by no means signifies that the investigational urge of later scientists was not intimately connected with their religiosity. The tenet that the scientific experience of truth in any sense is "theoria," i.e., a view of God, might be said sincerely about the best of them. The search for knowledge without regard to its applicability for use has been "an essential trait of man through the centuries, a sign of his higher origin."[3]

Physics has always been in close touch with its fellow sciences: astronomy, chemistry, and mineralogy. The boundaries separating it from them are marked only by rather superficial differences, characterized especially by the dissimilarity in apparatus; consequently the fields have frequently overlapped. In the seventeenth century, and even later, it was not rare to find an astronomer, physicist, and chemist united in a single person. Robert Boyle (1627-1691) and Edme Mariotte (1620-1684), who will be mentioned later, were primarily chemists, and this was also true of Henry Cavendish (1731-1810), Antoine Laurent Lavoisier (1743-1794) and Humphry Davy (1778-1829). Physics and chemistry have participated equally in the creation of the atomic concept. As a result of the work of Svante Arrhenius (1859-1927), Jacobus Henricus van't Hoff (1852-1911), Wilhelm Ostwald (1853-1932), and Walter Nernst (1864-1941), physical chemistry emerged as a

---

[3]R. Jaspers, *Die Idee der Universität,* Berlin, 1946.

4

separate science at the end of the nineteenth century. After a long interval, the physicists, in the twentieth century, began to concern themselves once again with the theory of crystals, which had otherwise been left to the mineralogists.

The connection between physics and mathematics is almost more intimate. The latter is the mental tool of the physicist. It alone enables him to express the natural laws in a final, precise, and teachable form; it alone makes possible their application to more complicated processes. For instance, logarithms, which were invented soon after 1610 by Jost Bürgi (1552-1632) and independently of him by John Napier (1550-1617) and Henry Briggs (1556?-1630), received one of their first applications in Keplers astronomical computations. Likewise, the progress of physics in later years, especially in mechanics, was most intimately allied with the concurrent advances in mathematics (see Chapter II). More than once, problems posed by physics have directly initiated mathematical advances.

The relation of physics to philosophy is quite special. At the opening of the period being considered here, physics also occupied the attention of some men who are known to us primarily as philosophers. Examples are Leibniz and Descartes, who, it is true, fundamentally rejected the Galilean method of investigation. Even Kant was active in science; the best known of his physical achievements are his cosmological ideas regarding the origin of the planetary system. dAlembert is better known as one of the leaders in the French "enlightenment" than for his accomplishments in mechanics. Later, the relations were reversed; physicists and chemists wrote on philosophy. Pertinent examples are Helmholtz, Mach, and Poincaré. They treated questions mostly related to the theory of perception, which, of all the philosophic disciplines, appealed most strongly to them. The author begs his readers indulgence if he doubts that all these scientists-philosophers possessed the philosophical training essential to a successful handling of their subsidiary field. However, there is no doubt that the advances of the natural sciences furnished a powerful impetus on all philosophers of eminence. The best known example is the influence of Newton on Kant. In the nineteenth century there appeared an all too justified opposition by the scientists to the "identity philosophy" of Hegel, which denied the right of existence to all empirical science. Unfortunately, this opposition was often extended to the entire field of philosophy, and in fact to all theory whatsoever in natural science. For example, J. R. Mayer, the champion of the energy principle, suffered from such attacks because of the highly speculative complexion of his writings. In fact, such objections were raised even against Helmholtz, when he first issued his famous treatise on the conservation of energy.

The relations between physics and technology are quite clear. The latter for the most part is applied physics, and its advances usually have followed closely on the heels of the progress in physics itself. However, technology has developed some ideas of its own which have proved to be of value for physics. Instances of such contributions are the steam engine by James Watt in 1770, and the setting up of the dynamo-electrical principle for the generator by Werner von Siemens in 1867. Above all, technology, in ever-increasing measure, has enlarged the experimental possibilities of physics. It would be utterly impossible to fit up a modern physics research institute without the extensive aid of technology.

Priority polemics constitute an unfortunate chapter in the history of every science. Even today it is difficult to decide such questions because every tolerably noteworthy advance is published in a periodical and the scientific press is only passably well organized. How much worse were the conditions when the news of the results of investigations could be spread only by books or in letters! There were no scientific journals prior to the middle of the seventeenth century. The Royal Society, founded in 1662, began to issue its Transactions in 1664. This example was followed, at considerable intervals, by the other scientific organizations and by the many academies founded around 1700. Thus, a system of sorts came gradually into the business of publishing results. Priority matters will not be given much attention in this book. From our standpoint it is much less important that the gas law named after Robert Boyle and Edme Mariotte actually was read out of Boyles measurements by his otherwise unknown pupil Richard Townley, than that the existence of this law was recognized about 1662 – but of course not by everybody; it had to be discovered again by Mariotte, independently of Boyle.

However, it is invariably true, no matter what the period, that if an investigator publishes a fundamentally new fact, sooner or later voices will be heard claiming priority either for themselves or for a third party, because it is alleged that they "really" had made the discovery earlier. Sometimes such claims possess a measure of justice. Cases can be cited in which a certain discovery was "in the air" and actually was made by several entirely independent workers because events had reached the point where the discovery was the natural next step (see Boyle and Mariotte). Rutherford[4] states that it is a far rarer case for a scientific discovery to be made without the apposite mental preparation of the world of science. In addition, such claims should be received with skepticism. Quite often vaguely expressed notions are subsequently embellished with a clear interpretation derived entirely from the work of some one else. Sometimes a man has had an idea or has made an observation the significance and importance of which are not appreciated until they are pointed out at a later time by another. A discovery should be dated only from that time at which it was so clearly and definitely stated that it had a distinct effect on further progress. If it is really announced in this form, then petty criticism should not be leveled against the text of the announcement, because it does not contain every incidental point in perfect order. Perfection has never been conferred on any mortal.

The history of nations and peoples records only such events and persons as have some kind of significance. Likewise, the

---

[4]Lord Rutherford, *Background of Modern Science*, Cambridge, 1938, p. 55.

history of a science can include only certain memorable points of investigations and those who participated in them. Thousands must remain unmentioned who, since the seventeenth century, set physics on the move and have devoted themselves to this science, many because of pure idealism and sometimes at the cost of self-sacrifice. However, their labors were neither superfluous nor in vain. The silent collaboration of these many unsung workers was required to produce the necessary profusion of observations and computations and they insured the continuity of progress. It was only the variety of interests and talents that prevented the researches from being confined within a few restricted directions. The activities of these many now forgotten workers constituted and still provide the indispensable preliminary setting in which outstanding accomplishments can be produced, including even the strokes of genius. Since the end of the seventeenth century, physics has been a highly cooperative effort. This, too, is an historical fact.

The question is often raised as to the objectivity, the truth of scientific knowledge. It is by no means accepted without doubt. There have been and still are perception-theoretical movements – and these recently were widely disseminated through political propaganda – which, basing their case on the human fortuity in the origin of all knowledge and the frequent change in physical views and theories, draw the conclusion that the whole is dependent on all possible environmental factors, mental or even biological, and therefore completely determined by time and convention. As a matter of fact, physics never has had a completely rounded-off form that lasted through all periods of its history; furthermore, it never can have, because the finiteness of its content will always be opposed by the infinite abundance of possible observations. Yet, it itself furnishes proof of its objective truth, proof that has overwhelming power of conviction. A study of the history of this science reveals repeatedly that two trains of physical thought, e.g., optics and thermodynamics (Chapter XIII) or the wave theory of X rays and the atomic theory of crystals

(Chapter XII), pursued up to then by different sets of workers, who were quite independent of each other, unexpectedly meet and fit together with no compulsion. Whoever has been privileged to live through such an extremely surprising event, even at a considerable distance, or, at least, to survey it after it has occurred, can no longer retain any doubt that the confluent theories certainly contain, if not complete truth, a substantial core of objective truth that is devoid of human embellishment. The ideal of a history of physics must be to set forth as clearly as it can such momentous events.

1 MEASUREMENT OF TIME

The measurement of time is one of the most important problems of every science that deals with events occurring in space and time. Why is this so?

In any case, Kant was correct in declaring that time is an idea inculcated in the human intellect. This concept is continuous, and in common with all continua it does not contain its measure within itself. Hence, in order to measure time, it is necessary to introduce a measuring system into it.[1] Intervals of time can be determined arbitrarily as, for instance, by tapping on a table and counting the strokes. If, then, the number of such time intervals coinciding with an event is stated, the time involved in a succession of events can be expressed by a series of numbers.

Obviously, any such method of dividing time into intervals must fail to meet the needs of even a great many events of daily life. For instance, a railroad time schedule could not be set up on this basis, since the running of locomotives obviously is governed by certain laws of nature, and the method of dividing time just suggested bears no relation to these laws. Consequently, the objective of measuring time must be a relationship with natural laws, and certainly in order to meet the demands of science, this connection must be such as to permit the formulation of the natural laws in the simplest

---

[1] A chain carries its own measure within itself; for instance, its links can be numbered. Nothing similar can be done in the case of a perfectly uniform thread. To determine the length, a rule must be placed alongside the thread, and the dividing marks transferred to it.

possible form.

Close examination reveals that this thought was the basis of even the ancient sand glasses and water clocks. It was established that a process, such as the passage of a given amount of sand or water through a certain opening, always took the same length of time. Experience had to decide how well such a fact would meet the particular need. However, such timekeepers stop after the fluid has run through; it is necessary to intervene in order to set them going again, and this operation interferes with the process of measuring the time. The same fault is inherent in the weight-driven clocks widely used in the Middle Ages. Their action depended on the fall of a weight, slowed down by an airbrake. It is likewise a defect of the simple pendulum, if it, following Galilei's example, is set in motion and the period of the swings is then used as a measure of time. Nonetheless, it was a pregnant advance when he recognized that the period is independent of the amplitude of the swings, even though, contrary to Galilei's belief, this rule is approximately true only for small amplitudes.

The decisive step, which made the clock possible, in its modern sense, was due to Christian Huygens (1629-1695), who was the first to recognize the ring of Saturn as such, and whose contributions to physics will be discussed later. In 1657 he introduced the principle of feed-back[2] a term now used in connection with the discovery (1906) by E. Ruhmer[3] of a method of producing electrical vascillations.

Fundamentally, all clocks consist of three essential parts. First, there is a swinger, usually in the form of a pendulum or balance, whose period supplies the actual measure of time. However, if new energy of motion is not continuously sup-

---

[2]Huygens obtained a patent on pendulum clocks from the States-General on June 16, 1657; his book, *Horologium*, appeared in 1658.

[3]E. Ruhmer's invention dealt with the arc transmitter. The feed-back was introduced in 1913 by DeForest, and almost at the same time by A. Meissner, for the vacuum tube circuits, which are far more important today.

plied, the oscillations must gradually die away because of the unavoidable frictional resistances. Accordingly, a second essential constituent is a source of energy, which stores energy in the form say of the elastic energy of a wound spring, or as the potential energy of a raised weight. In some of the newer forms of timepieces, these reservoirs consist of an electric battery. The third and chief of these essential parts is the apparatus, which transmits this energy to the swinger; it must do this in such manner that the latters period is not disturbed, and, in addition, the swinger itself must determine the instant at which the energy is imparted. This is the essence of the feed-back or escapement, which appeared first in the Huygens timepieces, both those with pendulums and those with balance wheels. Of course, all types of construction require that the energy source be wound up from time to time. However, this intervention, in principle, does not disturb the running of the timepiece. Therefore, it can be said that a clock or watch of this kind essentially measures time for unlimited periods.

Technology has, of course, greatly improved timepieces. The standards of accuracy, which are expected in even a moderately good watch of today, were impossible of attainment in Huygens time. However, the only profound change did not come until 1929 when W. A. Marrison discovered the quartz clock, which was developed further by A. Scheibe and U. Adelsberger. The swinger of this device is a quartz rod, which makes about 100,000 oscillations per second, and by virtue of the piezoelectric properties of the quartz is electrically coupled back with an electric battery.

In order to adapt the measuring of time to daily life, the timepieces have up to now been standardized against the rotation of the earth with respect to the fixed stars. A sidereal day is represented by two passages of the same star through the meridian, and the mean solar day, which is divided into 24 hours of 60 minutes each, which in turn consist of 60 seconds each, is 3465 longer than the sidereal day. The actual solar day, measured between two successive crossings of the meridian by the sun, varies in length throughout the year. Hence, all sun clocks show deviations up to one-quarter of an hour in comparison with correct mechanical clocks. The physics based on this means of measuring time explains this difference as being due to the deviations of the earths orbit from a true circle and to the inclination of the ecliptic.[4] A physics, which would attempt to measure time on the basis of the actual solar day, would have to deal with the awkward problem that all artificial clocks uniformly show annual deviations in their running.

It is obviously pure hypothesis to assume that the period of rotation of the earth is suitable for standardizing timepieces, in other words, that the rotational velocity of the earth is constant within periods measured by other good timekeepers. The test can be made in two ways. The time as given by two agreeing excellent quartz clocks seems to indicate variations in the time of rotation amounting to thousandths of a second. However, much more certain information has been secured from the comparison with the movements of the moon and the inner planets. These observations show that the time as read from the rotation of the earth compared with that required to understand these movements physically in the course of the past two centuries has varied over a range of as much as 30 seconds too early or too late.[5] In accord with the foregoing objective definition of measuring time, the time as given by the planetary clock must be chosen as correct.

This discussion has omitted any consideration of the fact that the location of every timepiece travels with the earth around the sun, and because of the earths rotation, the clock also moves around the axis of the earth. The relativity theory states that this actually introduces the necessity for a correction, but it also establishes by computation that the

---

[4]This means that the axis of the earth is not perpendicular to the plane of its orbit, but is inclined at an angle of about 23.5° with the Manual.

[5]Meyermann, Die Schwankungen unseres Zeitmasse" in *Ergeb. der exacten Naturwissenshaften* 7, 98 (1928).

correction is not significant as long as measurements cannot be made with more accuracy than at present.

2 MECHANICS

In the beginning was mechanics. As stated, the theory of equilibrium or statics extends far back into antiquity. It was brought into being by the practical importance of the lever, screw, block-and-tackle, and inclined plane as aids in heavy manual tasks. Such concepts as specific gravity and center of gravity were developed by the Greeks. The calculation of the center of gravity of a body of specified shape was a favorite mathematical exercise which required considerable skill as long as differential calculus was not available. Ancient statics reached its peak in the law of virtual displacements: multiply every force by the length of the path which the point of application traverses, provided a definite motion is produced. This motion will not ensue if the sum of these products (each given its appropriate sign) equals zero. Forces are measured here through weights; consequently, actions of gravity are always involved. The familiar law of the lever is a special case as is Archimedes' principle, which states that every solid body immersed in a liquid is buoyed up by force equal to the weight of the displaced liquid. The millennia before 1600 produced this knowledge at the cost of great labor. The last in the series of its creators was Simon Stevinus (1548-1620), who studied the equilibrium on the inclined plane in a brilliant, intuitive manner and thereby deduced the resolution of a force into components, i.e., he discovered the principle of the parallelogram of forces. The remainder of the mechanics taught by Aristotle, held to be incontrovertible truth through the entire scholastic period, proved to be nothing but the greatest of all the handicaps, which the budding science of the sixteenth century had to overcome.

The founding of the actual science of motion, i.e., dynamics, was due to Galileo Galilei (1564-1642); it was further developed by Christian Huygens; and brought to a certain degree of completion by Isaac Newton (1642-1727) in whose honor it is now known as Newtonian dynamics. Galileis studies of falling bodies commenced soon after 1589; his chief work on mechanics *Discorsi e Dimonstrazioni matematiche intorno a due nuove Scienze attenenti alia Mecanica & Moxnmenti locali* was published in 1638; Newtons *Philosophiae naturalis principia* appeared in 1687. Hence the creation period of dynamics was just about a century in length.

The result of this magnificent achievement of the human mind is contained in two laws: The product of the mass of a mass point times its acceleration is equal to the force acting on it. (Acceleration and force are directed quantities, i.e., vectors, and the law assumes, among other things, the same direction for both of them.) The second law is that of the equality of action and reaction: The forces exerted by masses on each other are equal in magnitude but opposed as to direction.

These statements need analysis. As to acceleration, it had been cleared up, in essence, by Galilei when, with primitive mathematical tools, he studied the concept of variable velocity. Newton, who had available the calculus invented by him and also by Gottfried Wilhelm von Leibniz (1646-1716), was able to lighten the task for himself. Acceleration is the change in velocity per unit time, the derivative of the velocity with respect to time, and hence the second derivative, with respect to time, of the radius vector drawn from a chosen starting point to the place at which the mass point is located. If the location and the elapsed time are known, the velocity and acceleration are therefore clearly defined. The first law gives consequently a second order differential equation for the location as a function of time; its integration determines the path and the velocity with which it will be traversed. When no force is acting, the acceleration is zero, the motion is in a straight line with constant velocity, in conformity with the principle of inertia.

The second law states the meaning of mass and inert mass" If two masses mutually accelerate each other, the extents of the effects are inversely proportional to the masses. This is likewise true, in case the motion is from rest, for the velocities attained in equal times and for the distances covered. Geometric measurement of the distance therefore makes it possible to refer every mass back to an arbitrarily chosen unit mass. Since the accelerations are in opposite directions, the sum of the products of the mass times the velocity remains unaltered, namely, equal to zero, provided both masses started from rest. As this product is defined as impulse, the foregoing laws can be restated in the form preferred today:

1. The force is equal to the change in impulse per unit time.[1]

2. In a system that is not influenced from without, and consisting of two, or even any desired number of masses, the total impulse is constant. (Law of the conservation of impulses.)

It is implicit in these statements that the forces exerted by two bodies on each other are not disturbed by a third body and that the mass is an unchangeable characteristic of the bodies. The latter assumption has always been an *a priori* postulate in mechanics, because no changes in the mass were ever revealed by weighings. Similarly, one of the most important facts learned in chemistry, which was developing into a science in the eighteenth century, was that the total mass of the reacting substances remains constant during chemical reactions. Antoine Laurent Lavoisier rendered particular service in this respect. A series of especially careful weighings, made in the years 1895 to 1906 by Hans Landolt (1831-1910), substantiated this belief. Nevertheless, today the constancy

---

[1] Even Newton used this formulation.

of mass is regarded as only an approximation that admittedly is fully adequate to the needs of mechanics, chemistry, and many branches of physics.

In the experiments, which provided the basis for this result, the forces were measured by means of weights, a long approved practice that is still in vogue. If the weights did not act perpendicularly downwards, the cords holding them were drawn over drums. Hence, the concept of force was really quite well established by experiment, and therefore it might well be thought to have been divested of every thing of a secret or metaphysical nature. But the seventeenth and eighteenth centuries were by no means so logical. The fact that the abstract meaning of the word force" was not entirely clarified led to confusion upon confusion. Since every conscious employment of force by man is preceded by an act of will, something deeper was sought within the physical notion of force. This mysterious something in the case of gravity, for example, was thought to be an innate tendency of bodies to unite with others of their own kind. It is difficult for us moderns to comprehend this standpoint. How generally it was accepted even by leading minds of the time is shown by the famous dispute over the natural measure of force between the Cartesians and Leibniz and his followers. One party took this to be the impulse produced in a given time by the force, the other side believed it to be what is now known as kinetic energy, which formerly was often called vital force. Newton was not able to take a definite stand on this matter. Although even dAlembert (1717-1783) labeled the endless discussion simply a battle of words, the concept of force in many minds, nevertheless, retained something of its mystical nature up to 1874, when Gustav Robert Kirchhoff (1824-1887) uttered the redeeming word in the first sentence of his Lectures on Mechanics. Mechanics is the science of motion; its task is to describe completely and in the simplest manner the motions occurring in nature." Accordingly, it is merely a matter of treating the vector denoting force as a function of the location of the mass point or the time, or even of both. The velocity can also be a determinant, in frictional forces, for instance. The integration of the Newtonian equation of motion then becomes a purely mathematical problem, whose solution provides the answer to every justifiable question concerning motion. Physics cannot and need not do more than this. If the reader finds something of causal explanation lacking in the word describe," he should note that the explanation of a natural event can consist only of bringing it into relationship with other occurrences by means of known natural laws, i.e., by describing a complex of related events as a whole. This fact has now been generally accepted and prevails in other fields as well as in mechanics.

A second series of important developments came in the same period. In 1643 Evangelista Torricelli (1608-1647), prompted by an experiment performed with a suction pump by Galilei, invented the mercury barometer. Blaise Pascal (1623-1662) in 1648 instructed his brother-in-law Perier to compare the height of the mercury column on the Puy de Dôme and at Clermont (a difference in elevation of about 1000 meters). Otto von Guericke (1602-1686) invented the air pump and with its aid cleared up the nature of atmospheric pressure by means of many impressive experiments.[2] It has already been pointed out in the Introduction that the Boyle-Mariotte law stating the relation between pressure and volume of the air was known by 1662. At that time, other gases[3] were not available since hydrogen was not discovered by Henry Cavendish until 1766; oxygen, by Karl W. Scheele (1742-1786), in 1769; and nitrogen in 1772, by Daniel Rutherford (1749-1819). In 1676, Robert Hooke (1635-1703), a contemporary of Pascal, discovered the proportionality in simple cases between deformation and stress in solids.

---

[2]The Magdeburg hemispheres were demonstrated in 1656. However, Guericke did not write a comprehensive account of his experiments until 1663; it was published in 1672 as Experimenta Nova (ut vocantur) Magdeburgica de Vacuo Spatio.

[3]The word gas is found about 1640 in the writings of the Dutch chemist-physician J. B. van Helmont (1577-1644); presumably, it came from the word chaos, employed by Paracelsus for mixtures of airs.

Thus, around 1700, were laid the physical foundations on which the next century and a half could build the magnificent structure of mechanics. Its completeness is characterized by the fact that this development lay predominantly in the hands of the mathematicians. The French took the leading part in this movement during the eighteenth century. In fact, Newton's ideas were propagated first in France, not only among the men of science, but the Enlightenment' carried them into far wider circles. This is a model example of the influence of physics on the general mental growth, and therefore also on political development. Special mention of the following is merited: Daniel Bernoulli (1700-1782), Leonhard Euler (1707-1783), who studied systems of several mass points, solid bodies, and hydrodynamics; Jean Lerond dAlembert, the author of the principle that bears his name and which replaces the equations of motion; Joseph Louis Lagrange (1736-1813), who gave these differential equations a form especially suited to more complicated cases; Pierre Simon Marquis de Laplace (1749-1827), whose five-volume Mécanique celeste, which appeared in 1800, contains much more than its title implies, namely, among others, a theory of liquid waves and capillarity. Thus the highest flowering of analytical mechanics was reached. Mention should be made also of: Louis Poinsot (1777-1859) to whom is due the completion of the theory of the rigid body; Gaspard Gustave Coriolis (1792-1843), who analyzed the effect, for instance, of the earths rotation on the events that took place on this planet; Augustin Louis Cauchy (1789-1857), who, in 1822, contributed the most general mathematical formulation of the exceedingly important concepts of elastic strain and deformation, and by using Hookes law, gave the mechanics of deformable bodies its final form; William Rowan Hamilton (1805-1865), who set up the principle of least action, which will be discussed presently; Karl Gustav Jacob Jacobi (1804-1851), who invented the method of the Hamilton-Jacobi differential equation for systems of several bodies. The studies of Jean Leon Poiseuille (1799-1869) on the internal friction of liquids and gases (1846-47), and the Helmholtz vortex laws (1858) can be considered as essentially closing this epoch, even though subsequent eminent investigators, especially Lord Rayleigh (1842-1919), Osborne Reynolds (1842-1912), and L. Prandtl still further advanced the dynamics of frictional liquids and gases. Such studies are still being carried on, particularly for purposes relating to the construction of water and air craft. The difference between orderly laminar and disorderly turbulent flow plays a part in this. If, however, experimental studies are also added, sometimes with enormous technical expenditure, this is done solely because the corresponding problems cannot be solved by present-day mathematics, or only with the expenditure of an inordinate amount of time. Nobody expects these studies to yield results that would go beyond the Newtonian foundations.

Only two results from the wealth of post-Newtonian development will be emphasized here. From Eulers time on, the mathematicians had set up variation principles, which were equivalent to the equations of motion, in fact, they contained the latter within themselves. A form of a principle of this type, which bears his name, was enthusiastically promulgated by Pierre Louis Maupertuis (1698-1759), but Lagrange was the first to state it correctly. The best known of these is Hamiltons *principle of least action*, which in 1886 was applied to a whole series of nonmechanical processes by Hermann von Helmholtz (1821-1894). Max Planck (1858-1947) regarded this as the most comprehensive of all natural laws. It deals with a time integral, to be formed between two fixed points of time with respect to the difference of the potential and kinetic energy, and states that for the actual motion this integral is smaller than for any other conceivable one that leads from the same initial to the same final condition. When such principles were brought out in the eighteenth century, they caused a tremendous sensation. The differential equations of motion determine what happens at a given instant from the immediately preceding motion, in conformity with the causal concept of nature. In these principles, on the contrary, the entire motion over a finite period of time is taken into account all at once, as though the future plays a part in determining the present. Accordingly, a teleological factor seemed to have been introduced into physics, and certain enthusiasts even went so far as to imagine that they were being given here a glimpse into the world plan set up by the Creator, Who had ordained that the values appearing in these principles should be kept as small as possible. The Leibniz idea of "the best of all possible worlds" smacks of this notion.

Of course, a mathematical error was at the bottom of this doctrine. Later critical studies revealed that although these quantities always have an extreme value for the real motion, the value is by no means invariably a minimum. Furthermore, it soon became evident that variation principles can be set up for differential equations other than those pertaining to mechanics. Consequently, the principle of least action and all similar ideas were put back into their proper position as highly valuable mathematical aids.

This could be an appropriate place to mention a second, and far more important point, namely, the law of the conservation of energy, which had had a history within mechanics even before it emerged from this province to become a universal law. However, it will be discussed in Chapter VIII.

R. W. Hamilton, who also contributed to the development of geometric optics, pointed out the mathematical similarity between this discipline and mechanics. A light ray and the path of a mass point correspond so that it must be possible to recombine the paths of all of the mass points which issue from a point with the same velocity into a "focus" and thus mechanically produce "optical" representation. Of course, this could not be accomplished until the discovery of electrons, i.e., of particles in which the action of gravity can be completely overshadowed by electrical forces. However, the electron microscope, at least in its electrostatic form,[4] is the direct application of the Hamiltonian concept.

---

[4]There is also a magnetic model.

The relativity theory, formulated in 1905 by Albert Einstein, does not greatly alter the dynamics of the mass point, as was shown by Planck in 1906. (Einsteins fundamental work is incorrect in this regard.) A distinctive feature is the inclusion of a universal constant, whose mechanical significance had hitherto been unrecognized, namely, the velocity of light in empty space. The proposition that force equals change of impulse per unit time is preserved, likewise the conservation of impulse in a closed system. Just as before, this gives rise to the energy law; but now the relations between impulse and energy change as the velocity changes. Although this change is noticeable only for velocities approaching that of light, nevertheless, in this region, impulse and energy increase without limit, with the result that no object can ever reach the velocity of light.[5] The latter is the unattainable upper limit of all corpuscular velocities. Electron velocities up to 99 per cent that of light and higher have been found in radioactive atomic disintegrations, velocities exceeding that of light have never been established experimentally. The validity of the relativity formula for impulse was established by numerous measurements[6] (1906-1910) of the deflection of fast electrons carried out by Walter Kaufmann (1871-1947), Alfred Heinrich Bucherer (1863-1927), Charles Eugene Guye (1866-1942), and Simon Ratnowsky (1884-1945).

The change in the mass concept, which this theory forces on the physicist, is fundamentally of still greater importance. As Einstein demonstrated from it in 1905, every addition of internal energy must increase the mass, and by an amount that is obtained by dividing the energy, measured in mechanical units, by the square of the velocity of light. However, because of the magnitude of this velocity (3×10^{10} cm/sec), the changes are insignificant in all processes which are desig-

---

[5] EDITOR'S NOTE: The way this is explained requires clarification to avoid confusion and misconceptions; see Editor's Note at the end of this chapter.

[6] EDITOR'S NOTE: What was established by those measurements is discussed in the Editor's Note at the end of this chapter.

nated as mechanical, electrical, or thermal. No change in the total mass of the reacting substances can be observed even in the most vigorous chemical changes that have the greatest heats of reaction. In nuclear physics, however, this law of the inertia of energy acquires an enormous significance (see Chapter X).

What does mechanics accomplish? Exceptionally much. It provides the basis for every technical construction, in so far as the latter is mechanical, and it thus enters intimately into daily life. It finds application in the biological sciences; for instance, as mechanics of the bodily movements or of hearing. It contains the theory of the deformation of solids that are subjected to elastic stress, of flow in liquids and gases, and furthermore of the elastic vibrations and waves that are possible in all such bodies, i.e., of the whole field of acoustics, to the extent that the latter is physical in nature. It has, to emphasize a particular case, led to a theory of covibration, whose significance goes far beyond the province of mechanics and which, for instance, is basic to the understanding of electrical oscillations. It describes, in agreement with all observations, the motions of masses whose weights range from that of the fixed stars ($10^{32} - 10^{33}$ grams) to that of ultramicroscopic particles (10^{-18} gram). In fact, it has confirmed, in part, the experimental data on the motion of molecules, atoms, and the still smaller elementary particles (electrons, etc.). Consequently it became the basis of the kinetic theory of gases and of the Boltzmann-Gibbs formulation of physical statistics. It combines all these elements into a structure of majestic architecture and imposing beauty. Hence, it is not surprising that for many years mechanics was regarded as equivalent to the whole of physics and accordingly the purpose of the latter was viewed frankly as an effort to relate all processes back to mechanics. Even after it was realized, around 1900, that this could not be done for electrodynamics, many erroneously still considered mechanics as ranking above experience, like mathematics, for instance. The shock was therefore all the greater, when, from 1900 on, the validity of the quantum theory became increasingly evident in ever-widening areas. But even where it displaces mechanics, this theory retains unchanged two of the latters laws: the conservation of energy and the conservation of impulse.

Acoustics, however, is one branch of mechanics that developed rather independently, particularly in its earlier stages. Even the ancients knew that pure tones, in contrast to noises, are due to periodic vibrations of the source of the sound. Pythagoras (582?-500? B.C.) knew, in addition, perhaps from Egyptian sources, that strings, which are tuned in harmonic intervals of octaves, fifths, etc., have lengths, which, other conditions being constant, are in the ratio of 1:2:3 and so on. The deep impression that this discovery made came from the great stress which the Pythagoreans placed on number in their general view of the world. Organs were widely distributed as early as the ninth century of the Christian era, and the builders must have known the corresponding facts about organ pipes. However, other than this, the knowledge of acoustics apparently took no part in the great advances experienced by the musical art in the two millennia after Pythagoras. Again, it was Galilei who provided the decisive impulse that promoted further development. In his *Discorsi* of 1638 (see p. 18), he declared that the vibration frequency is the physical correlate of the sensation of pitch; he regarded the relation of the vibration frequencies as determinants of the relative heights of two tones, and he also showed how the vibration frequency of a string depends on its length, the tension to which it is subjected, and its mass. He observed and explained the excitation of vibration through resonance, and he also particularly recognized the existence of stationary waves, the latter on the surface of water in vessels, which he had caused to emit notes by rubbing. At about the same time, namely in 1636, his former pupil Marin Mersenne (1588-1648) advanced somewhat farther. He made the first absolute measurement of vibration frequencies and of the speed of sound in air. In addition, he contributed the observation that a string usually emits its harmonic overtones along with the fundamental. Joseph Sauveur (1653-1716) did the same for organ pipes; he was acquainted with the properties of vibrations and determined the position of nodes and loops on vibrating strings by means of paper riders, a method still in use.

The fact that sound, in contrast to light, is not transmitted through an evacuated space was experimentally demonstrated by Otto von Guericke. The relation of the speed of sound to the compressibility and density of the atmosphere was calculated for the first time by Newton in his *Principia*, although his formula did not agree with experiment until Laplace, in 1826, replaced the isothermal by the adiabatic compressibility. The mathematical treatment of mechanics in the eighteenth century also benefited acoustics. However, the latter produced no other eminent experimenter until the advent of Ernst Friedrich Chladni (1756-1824), the Father of Modern Acoustics. In 1802, he, among others, compared the longer known transverse vibrations of strings and rods with longitudinal and torsional vibrations; he made visible, by means of sand figures which now bear his name, the nodal lines of vibrating plates; he also measured the velocity of sound in gases other than air. Despite the direct observation (1762) by Benjamin Franklin (1706-1790), doubt persisted for many years regarding the transmission of sound through liquids, because they were supposed to be incompressible. However, incontrovertible proof of this fact was furnished in 1827 by Jean Daniel Colladon (1802-1893) and Jacob Franz Sturm (1803-1855) who found the speed of sound in Lake Geneva to be 1.435×10^5 cm/sec.

During the further course of the nineteenth century, physical acoustics became increasingly a part of the field of elastic waves. The ideas of interference, diffraction, and dispersion at obstacles, were carried over into sound from optics. Dopplers principle, which originated in 1842 as an optical idea (Chapter VI), received its first confirmation in the changes in pitch that are heard when whistling locomotives go by. Fourier analysis, which was originally invented to deal with problems in the conduction of heat (Chapter VII), experienced a triumph when it was applied to sound waves, especially since the resolution of any periodic vibration into sinusoidal vibrations corresponds to a direct psychological reality; the ear is capable of hearing the sinusoidal vibrations separately, a fact established in 1843 by Georg Simon Ohm (1787-1854). In cases where this analysis is not possible, because of inadequate intensity or lack of practice, these vibrations nevertheless determine the timbre or quality of the mixture of tones, a fact emphasized especially by Helmholtz in his *Theory of Tone Sensations* (1862).

Great technical problems were presented to acoustics after Philipp Reis (1834-1874) and Alexander Graham Bell (1847-1922) invented the telephone in 1861 and 1875, respectively, and again after 1878 when David Elwood Hughes (1831-1900) materially improved the Reis microphone. The best possible reproduction of the human voice and musical sounds had become necessary. The importance of the new field of application, electro-acoustics, was further heightened by the transmission of sound by means of electrical waves, a fruit of World War I (1914-1918). The phonograph, invented in 1877 by Thomas Alva Edison (1847-1931), also falls into this category.

During this same war, Paul Langevin (1872-1946) found that quartz plates, if excited by piezoelectricity, could be used to produce sound waves in water, with vibrations of the order of 100,000 per second, i.e., far above the audible limit. This ultrasound was to be used in the detection of submerged submarines. These ultrasonic waves were subsequently used by physicists in studies of the properties of solid bodies, for measuring the velocity of sound in gases and liquids with respect to vibration frequency, and for various other purposes. Such waves also play a certain role in biological research.

In conclusion, mention should be made of another advance, which, though it was of a more public nature, nevertheless, had a marked effect on the whole of physics. On June 2, 1799, the Legislative Assembly in Paris adopted the kilogram as the standard unit of mass and the meter as the unit of length. These units, together with the much older unit of time, the second, form the basis of the cgs (centimeter/gram/second) system, to which modern physics relates all mechanical electrical, and magnetic units.

EDITOR'S NOTE

Two things von Laue wrote on p. 25 need clarification.

1. According to him, "impulse and energy increase without limit, with the result that no object can ever reach the velocity of light."

The way this is stated may lead to confusions and misconceptions. What increases without a limit and prevents an object from reaching the velocity of light is the object's increasing (relativistic) mass. Since mass is defined as the measure of the *resistance* an object offers to its acceleration, when an object is accelerated to velocities approaching that of light, it is an experimental fact that *its resistance increases* and approaches infinity. It is this increasing resistance (i.e., its increasing relativistic mass) that does not allow an object to travel as fast as light.[7]

---

[7]That is why, it cannot be stated that it is sufficient to say it is the objects energy that increases with its velocity, because the crucial experimental fact is the *increasing resistance* the object offers to its acceleration and the measure of this resistance is the objects mass. It

Max Born explicitly warned about the danger of improper understanding of mass in relativity:[8]

> In ordinary language the word mass denotes something like amount of substance or quantity of matter, these concepts themselves being defined no further... In physics, however, as we must very strongly emphasize, the word mass has no meaning other than... the measure of the resistance of a body to changes of velocity.

2. What also needs clarification is the statement: "The validity of the relativity formula for impulse was established by numerous measurements." What these measurements decisively confirmed is the relativistic increase of mass. For example, Bucherer measured the ratio of charge to mass (e/m) for β-ray electrons and showed that at high velocities, comparable to the velocity of light, the masses of the electrons depended on their velocities. This experiment allowed *only* two interpretations – that either e or m varies in the ratio e/m – and independent experiments[9] ruled out the interpretation that the electron charge decreases as its velocity increases. Therefore, the Bucherer experiment would be *impossible* if the mass of electrons did not increase as their velocities increase.

---

cannot be stated either that it is the energy which resists the acceleration because there is no such thing as energy as such; it is the energy of an object and it is the object that resists its acceleration.

[8]M. Born, *Einstein's Theory of Relativity* (Dover Publications, New York 1965), p. 33

[9]See W. G. V. Rosser, *Relativity and High Energy Physics* (Wykeham Publications, London 1969), p. 15

3 GRAVITATION AND ACTION AT A DISTANCE

The study of gravitation was intimately connected with the genesis of mechanics. Of course, the human mind has dealt with this phenomenon through all ages, from antiquity to the present, and no topic in physics, with the exception of atomistics, has given rise to as much speculation as the reasons for the force of gravity. What is actually known about it is due to those who limited themselves to the question: How does it act? Galilei went into this most extensively. He simply accepted as fact that bodies close to the earths surface receive a constant acceleration perpendicularly downward; this assumption sufficed for the derivation of his laws of falling bodies. Newtons famous hypotheses non fingo at the close of his *Principia* should also be recalled here. Both of them, however, laid great stress on the proof that all bodies experience the same acceleration, and tested this not only with freely falling bodies, but also by showing that the period of a pendulum is independent of the nature of the swinging body. The most thoroughly conceivable opposite would be to postulate that the force of gravity is proportional to a heavy mass that is different from the inert mass. The equality of both masses is one of the most prominent features of the theory of gravity.

The idea that gravitational attraction is not confined to proximity to the earth, but is a general property of all matter, and therefore is active also between the heavenly bodies, is likewise quite old. Indications of it can be found, for instance, in the writings of Nikolaus Copernicus and R. Hooke. More respect was accorded during the seventeenth century, and by some even until well into the eighteenth century, to the theories of the great philosopher René Descartes (1596-1650). He rejected a void as a *contradictio in adjecto* and thought of the interstellar space as filled with a whirling fluid which carried along with itself the planets afloat in it. Newton devoted a considerable part of his *Principia* to the hydrodynamic refutation of this theory. When tracing out the origin of the Newtonian law of universal gravitation (the force of attraction is proportional to both masses and inversely proportional to the square of the distance between them), it is essential to call attention to the following great trio: Tycho Brahe (1546-1601), who made particularly accurate, well-planned series of observations of the positions of the planets; Johannes Kepler (1571-1630), who from the Tychonic data, deduced the three laws that bear his name (elliptical orbits; equal areas swept out in equal times by the radius-vector; the squares of the times of revolution are proportional to the cubes of the great axes), and who, like some of his contemporaries, already sensed the decrease of the force with the square of the distance; and finally, Isaac Newton, who proved this law by quantitatively confirming it with respect to the acceleration on the earths surface, and that experienced by the moon, and, in addition, mathematically deduced from it the laws of Kepler and his own universal law of motion. The latter rightly bears his name. Although many before him, such as Robert Hooke, had spoken of universal attraction between all bodies, none had understood how to derive from it, by rigid mathematics, the laws of Kepler together with the slight deviations from them due to the mutual disturbances of the planets, to point out the anomalies in the motion of the moon, and to elucidate the connection with the ebb and flow of the tides. These accomplishments must have made a tremendous impression on his contemporaries. This is readily understandable because hitherto the positions of the planets had been considered to be the direct result of divine will. With a single stroke, this new knowledge demolished the structure of astrology, which, up to then, had been highly esteemed. Certainly, nothing in its previous history had contributed so powerfully to the establishment of respect for the young science of physics as did the Newtonian computation of the orbits of the planets. From then on, this science has been a great mental empire, which no other power may ignore with impunity.

The Newtonian law signified also a complete change in the conception of the systems of fixed stars. From Aristotle up to and including Kepler, with the exception, of course, of Giordano Bruno (see Chapter VI), the fixed stars were thought of either as being attached to the surface of a sphere, which had the sun as its center, or as being in a relatively thin spherical shell, beyond which there was thought to be absolutely nothing, not even space. Kepler himself had refused to regard the sun as merely one of many fixed stars. Now it became clear that the system of fixed stars does not represent something static, but instead forms a dynamic whole, held together by gravitation, but subject to internal motion, i.e., to the succession of the constellations. There immediately arose such questions as: "How far away are the various fixed stars? and Do the stars have a motion of their own in the celestial vault? These problems were attacked successfully only at a later period, but the concept of an unlimited universal space and the inclusion of the sun among the fixed stars was nevertheless a necessary deduction from the Newtonian law.

The law of gravitation contains a proportionality factor, the gravitational constant, which represents the force with which two masses of one gram each attract each other when they are one centimeter apart. Astronomy can compare the masses of various heavenly bodies with each other, but it cannot determine this constant. The requisite experiment was carried out by Henry Cavendish in 1798. He used a torsion balance, an instrument that had been employed as early as 1785 by Coulomb (see Chapter V) for electrical measurements. The value of the constant is 6.7×10^{-8} cm^3/gm-sec^2, from which the mass of the earth has been calculated to be 6×10^{27} grams.

In conjunction with the universal law of gravitation, it should be pointed out that in 1777 Joseph Lagrange defined potential, whose gradient yields the force of attraction; that Pierre Simon Marquis de Laplace derived for this function the coordinates of the partial differential equation ($\Delta\varphi = 0$), which bears his name, and that, in 1812, this expression was modified in the well-known manner by Simeon Denis Poisson (1781-1840) to apply to the interior of matter. These were important advances preparatory to the potential theory of electrostatics (see Chapter V). The Laplace-Poisson differential equation is the generalized expression of the Newtonian law of attraction. It follows from the latter, but also leads back to it, when applied to mass points (or homogeneous spheres).

The law of gravitation provided a firm foundation for theoretical astronomy, whose most important problem, the calculation of the perturbations of the planetary orbits because of the mutual attraction of the planets is, of course, still occupying the attention of astronomers and mathematicians. Some of the mathematical methods of mechanics have developed from this law. A milestone on this road was the *Mécanique celeste* of Laplace, which was published about 1800. The excellence of the law was illustrated most conspicuously by the discoveries of Neptune and Pluto, since even before these outermost planets were ever seen, their positions were computed from the perturbations of the orbits of the next inner planets. Neptune was found in 1846 by Johann Gottfried Galle (1812-1910) on the basis of calculations by Urbain Jean Joseph Leverrier (1811-1877), and Pluto in 1930, from the computations of Percival Lowell (1855-1916), at the observatory named after him at Flagstaff, Arizona. Only a minimal lack of agreement remained. Because of the deviation of the affected planetary orbit from the elliptical form, the perihelion, i.e., the position of the shortest distance of the planet from the sun, slowly rotates in the plane of the orbit. The perturbation theory explains this in quantitative agreement with the observations for all the planets except the innermost one, Mercury. In this instance, a deviation of 42 seconds of arc per century remained unaccounted for. In 1916 Einstein's general relativity theory (1913 on) explained this as being a result of the warping of space,[1] which according to this theory accompanies every gravitational field, but becomes appreciable only in the neighborhood of a tremendous body, such as the sun. The fact that a result of 42 seconds positive was obtained by a computation based on the mass of the sun, the gravitational constant, and the distance between Mercury and the sun, constitutes one of the two empirical supports of this brilliant law, which, however, cannot yet be regarded as fully verified.[2]

Newtons law of gravitation, if literally construed, assumes direct action at a distance. The possibility of such action has been doubted in all times, including Newtons; in fact, he himself was by no means unaware of this objection. The speculations, mentioned above, about the causes of gravitation, frequently originated in efforts to discover a transmission mechanism for gravity. Nevertheless, as a consequence of the amazing effect produced by the Newtonian discovery, the idea of action at a distance was carried over into other fields of physics. Certainly, one of the factors contributing to this spread was the fact that the simple and elegant mathematical potential theory can be derived from it. Whereas, close-up actions are employed exclusively in the mechanics of deformable bodies, the earliest theories of electrical and magnetic phenomena became distant action theories. The change did not occur until the second half of the nineteenth century, when the influence of Michael Faraday (1791-1867) and J. Clerk Maxwell (1831-1879) made itself felt. This movement prevailed after the discovery (1887) by Heinrich Hertz (1857-

---

[1]EDITOR'S NOTE: The precise expression is "warping of spacetime."

[2]EDITOR'S NOTE: Now Einstein's general relativity is fully verified, even on a daily basis through the GPS.

1894) of electrical waves that travel with the velocity of light. This brought about also some loss of standing to the distant action of gravitation, and matters really became serious when the special relativity theory (1905) stated that the velocity of light represents the maximum for the velocities of propagation of all physical actions. The general relativity theory (1913 on) also states that gravitational force[3] travels at the velocity of light and also postulates the existence of gravitational waves, whose production in observable strength of course cannot overcome the insuperable experimental difficulties. This theory retains the Newtonian gravitational law as an approximation.

A humorous demonstration of the stifling prestige of the Newtonian ideas is given by the fact that some scientists of the eighteenth century relegated meteors to the status of a legend, despite numerous observations, from antiquity on, that testified to their actual existence. Newtons unworthy followers believed that the chaotic descent of stones and masses of iron from heaven, was incompatible with the cosmic orderliness revealed by the master. It was not until 1794 that Ernst Friedrich , well versed in law, collected and made a critical study of most of the testimony, and from the widespread agreement between entirely independent accounts, concluded that the observers had seen real objects. When, in addition,

---

[3]EDITOR'S NOTE: The correct expression would be, for example, "gravitational influence travels at the velocity of light." The reason is that "In relativity there is no such thing as the force of gravity, for gravity is built into the structure of space-time, and exhibit itself in the curvature of space-time, i.e. in the non-vanishing of the Riemann tensor R_{ijkm}" (J. L. Synge, *Relativity: The General Theory* (Nord-Holand, Amsterdam 1960) p. 109). Not only is there no gravitational force in general relativity, but, most importantly, there is no gravitational force in Nature. It is an experimental fact that falling bodies *do not resist* their downwards acceleration (a falling accelerometer, for example, reads zero acceleration, i.e., zero resistance). This means that no gravitational force acts on them. A gravitational force would be required to accelerate a body downwards if and only if the body *resisted* its acceleration, because only then a gravitational force would be needed to *overcome* that resistance.

a great swarm of meteorites descended in 1803 in the vicinity of Laigle (Department of Orne), and Jean Baptiste Biot (1774-1862) was able to study this shower, the Paris Academic found itself forced to abandon its negative attitude. There were indeed more things between heaven and earth than are dreamt of in your philosophy.[4]

---

[4]Hamlet was certainly correct in this statement, but, in return, philosophy likewise contains things of which no trace can be found between heaven and earth. This witticism is ascribed to the Göttingen physicist Georg Christoph Lichtenberg (1742-1799).

4 Optics

Optics is not much younger than mechanics. The knowledge of the linear propagation of light and the concept "ray" extend back far into antiquity. The ancients likewise thought about reflection and refraction, and they were acquainted with methods of copying by means of concave mirrors and lenses. The position of the focal point, but also the inexactness of the reunion of the rays in the image of a point of light, were described by Roger Bacon (1214?-1294). Spectacles seem to have been invented around 1299 by Salvino degli Armati, a Florentine. The law concerning the direction of the reflected ray also is one of the oldest heritages of unknown origin. In contrast, the law of refraction can be credited to two men. The first, Willebrord Snell or Snellius (1591-1626), according to the testimony of Huygens, deduced the law from actual measurements and published it in a work that has been lost. The second, René Descartes (1596-1650) derived it from his corpuscular conception of light. Kepler was not quite correct in this matter; his formula holds only as an approximation for small angles of incidence, but it was good enough for him to develop a thoroughly usable theory of the telescope. Nevertheless, the laws of reflection and refraction provided a complete physical foundation for geometric optics, whose further development then came chiefly from mathematicians and instrument makers. Men such as William Rowan Hamilton and Karl Friedrich Gauss (1777-1855) participated but, despite the expenditure of much effort and acumen, the structure is not yet finished. Limits are set to its validity by the wave nature of light; they reveal themselves, in the case of the microscope, by the fact, known in 1874 by Ernst Abbe (1840-1905) and Hermann von Helmholtz, that with visible light there is no resolution of distances smaller than 10^{-5} cm. Of course, since this limit is less when ultraviolet light, whose wavelength is shorter, is employed, and with X-rays, as has been known since 1912, it is possible even to deal optically with the distances between the atoms of solid bodies. They are of the order of 10^{-8} cm (see Chapter XII).

The explanation of colors presented a particular difficulty to optics in its earliest period. The proof (1672) that white light is composed of light of various colors, and that consequently colored light is simpler in nature than white light, was Isaac Newtons second great accomplishment. Nothing illustrates its importance better than Goethes vehement protest (1791-92 and 1810) against it, which in the end goes back to the fact that the eye, unlike the ear, which harmonically analyzes the vibrations that are stimulating it, perceives white light as a unit. Newton was led to his studies with the prism by the chromatic error of optical instruments, a defect which he considered unavoidable. His design and construction of the reflecting telescope, which fundamentally steers clear of this fault, was a logical step. His successors also maintained this position, until 1753 when John Dollond (1706-1761) produced an achromatic telescope objective, in which the chromatic aberrations of two varieties of glass were mutually compensated. In 1800, Friedrich Wilhelm Herschel (1738-1822) found that the limits of the spectrum do not coincide with those of its visible region, but, on the contrary, less refrangible radiation, revealed by its heating effect, adjoins the red. The next year, Johann Wilhelm Ritter (1776-1810) and likewise William Hyde Wollaston (1766-1828) discovered the chemically active radiation beyond the violet.

The quantitative study of the continuum of different kinds of light, which the prism separated, presented a problem similar to that of the measurement of time (see Chapter I). The designations red, yellow, etc. were too loose to serve as divi-

sions within this band of light, and they were also too sub-
jective, since they differ from person to person. Hence, it
was a great advance, when, in 1814-15, Joseph von Fraunhofer (1787-1826), by introducing a collimator and telescope
before and behind a prism, discovered in the solar spectrum
the sharp, dark lines which now bear his name. He used them
as markers and immediately was able to define closely the
measurements of the refractive index by assigning a value of
it to each of these lines. This procedure is still used for some
technical purposes. However, the problem was not solved until 1821-22, when he discovered diffraction by a grating, and
Friedrich Magnus Schwerd (1792-1871) provided (1835) an
explanation of it based on the undulatory theory. From then
on, it has been possible to coordinate every variety of light
with wavelengths, which can be measured with relative ac-
curacies to within 10^{-7} cm from the constants of the grating
and the angle of diffraction. This marked the real beginning
of spectroscopy which has always had tremendous importance
for the entire field of science and technology. For instance,
wavelength measurements by F. Paschen (1865-1947) on the
lines of hydrogen and helium, and the precise determinations
of Rydbergs constant based on these data, became a decisive
verification of Bohrs atomic model (Chapter XIV).

Another question that was much discussed in the seventeenth century was: "Does light have a finite velocity?
Descartes denied, and Galilei affirmed, both without experimental basis for their respective opinions. In fact, the laboratory resources of the period were not adequate to deal
with this problem. However, in 1676, Glaus Römer (1644-
1710), using observations of the almost but not quite periodic
eclipses of one of the moons of Jupiter, came to his famous
conclusion that the velocity of light in empty space is about
3×10^{10} cm/sec. The aberration observations in 1728 by
James Bradley (1693-1762), despite the doubts of the Cartesians, provided the very welcome confirmation that it is 10^4
times greater than the velocity of the earth in its orbit. Laboratory determinations were not made until 1849 when Ar-

mand Fizeau (1819-1896) employed a rotating toothed wheel, and in 1862 Jean Bernard Leon Foucault (1819-1868) used a rotating mirror. These measurements were repeated by many, mostly by the same methods. The latest determination was made by Albert Abraham Michelson (1852-1931) in 1925 and 1926. He measured the time required for light to make a round trip between Mount Wilson and Mount Antonio in California, a total distance of 70 kilometers. His result was 2.99796×10^{10} cm/sec, with a probable error of 4×10^5 cm/sec.

The discovery of interference, diffraction, and polarization became decisive for the theory of light. The earliest of such observations were made by Francesco Maria Grimaldi (1618-1663) who in his posthumous book (1665) gives a detailed description of diffraction at a rod and a granting. Even after Newton repeated these experiments they remained without influence on the development. The same is true of the discovery by Robert Boyle of the colored rings exhibited by thin films, which now are called Newtons rings, because the latter was the first to recognize the relation between color and thickness of the layer. Newton accepted a corpuscular "emanation theory of light, though with some reservation, because he obviously put more weight on his experimental findings than on their elucidation. In the Introduction to his *Optics* he rejects the making of hypotheses just as firmly as at the close of the *Principia*. Nevertheless, his less brilliant successors tenaciously retained this theory, which survived even into the nineteenth century. For instance, it was supported by Jean Baptiste Biot, who was not convinced of the validity of the undulatory theory until quite late, even though he was permitted to witness its growth at close range.

Grimaldi only tentatively, and Hooke more decidedly, had considered a wave theory. However, the latter really dates from the "Traité de la lumiere, which Huygens presented to the Paris Academic in 1678, and which appeared in print in 1690. From the assumption of a longitudinal wave motion, he derives, by means of his envelope construction, rectilinear propagation, and the laws of reflection and refraction. The latter was applied not only to isotropic bodies, but also to calcite, whose double refraction he explained as due to the joint action of two wave fronts, of which one is a sphere, as in an isotropic body, and the other a revolution ellipsoid. There is nothing in Huygens book about the spectral resolution into colors.

In contrast to mechanics, there was practically no progress in the theory of light during the eighteenth century. Then there began the "heroic period of the undulatory theory, which extended from 1800 to about 1835; the development took place chiefly in England and France. Thomas Young (1773-1829) announced his idea of interference in 1801 and applied it in the familiar manner to Newtons rings. He thus became the first to obtain an approximate measure of wavelengths of light. Furthermore, he recognized the difference between coherent rays, coming from the same light source, and incoherent rays. He employed the idea in explaining diffraction, which he regarded as interference between the light going directly through the diffraction aperture and the boundary waves. This will be discussed later. Polarization was discovered by Etienne Louis Malus (1775-1812) in 1809. He believed it provided a refutation of the undulatory theory; but, actually, it is incompatible with the longitudinal waves of Huygens Traité. Thereupon, in 1811, Dominique Francois Arago (1786-1853) described the color phenomena that can be seen on crystals in polarized white light. As a consequence, Thomas Young found it necessary to declare his belief in the transversality of light waves, even though this idea definitely contradicted the usual views. In 1815, the brilliant Augustin Jean Fresnel (1788-1827) began his all too brief career of discovery. In addition to many new observations on diffraction and interference, he contributed the the theory of diffraction in the form of zonal construction, which firmly implanted the Huygens envelope principle in the interference idea. He and Arago in 1819 furnished the proof that polarized rays at right angles to each other do not interfere, a discovery which finally put the theory of transverse vibrations on a firm footing. His crystallooptics, which is still accepted, explained Aragos experiment, among other things. Finally, Fraunhofers diffraction studies (1821-22), which deviated from those of Fresnel and were simpler from the theoretical standpoint, came in this same period, which in a sense, was brought to a close by Schwerd in 1835 in his comprehensive work: "The diffraction phenomena analytically developed from the fundamental laws of the undulatory theory.

The idea of interference, that rays on meeting, in contrast to beams of corpuscular particles, do not necessarily strengthen each other, but rather can weaken each other to the point of complete extinction, has since then been one of the most valuable possessions of physics. Whenever the nature of a radiation is in doubt, an effort is made to produce interferences; if successful, the wave character is definitely established.

Light was thus proved to be a transverse wave motion. In the course of time, interference apparatus and experiments increased greatly in proportion to the advance in the art of experimentation, and they, in turn, contributed to greater accuracy of measuring. About 1834, Macedonio Melloni (1798-1854) showed that infrared rays behave as light does in reflection, refraction, and absorption experiments, and in 1846 Karl Hermann Knoblauch (1820-1895) by means of diffraction, interference, and polarization experiments, demonstrated that such radiation differs from light only with respect to its greater wavelengths. The new art of photography was applied to the shorter wavelength ultraviolet radiation. A still greater extension of the knowledge of the spectrum came in November, 1895, when Wilhelm Conrad Röntgen (1845-1923) announced his epoch-making discovery. Almost immediately (1896) it was concluded by Emil Wiechert (1861-1928) and George Gabriel Stokes (1819-1903) that Röntgen rays, judging from the manner in which they are produced, must be a type of radiation with particularly short wavelength. This conclusion was fully confirmed by the polarization studies of C. G. Barklas and the interference investigations by W. Friedrichs and P. Knipping (1883-1935) on the atomic space lattices of crystals. The wavelengths of X-rays range between 10^{-7} and 10^{-9} cm.

The theory of optics was also making progress. First of all, Young's idea of boundary waves in diffraction received an unexpected substantiation when L. George Gouy (1854-1926) in 1883 and Wilhelm Wien (1864-1928) in 1885 observed them directly in the light deflected at large angles; up to then, the observations had been confined to the immediate vicinity of the shadow boundaries. G. R. Kirchhoff, whose work on the mathematical wave theory was continued by A. Rubinowicz in 1917, put Fresnels brilliant idea of zone construction on a firm basis in that, for instance, he proved mathematically the identity of the Young and the Fresnel concepts of the diffraction process. All these studies were dependent on approximation methods, but in 1894, A. Sommerfeld successfully dealt with the diffraction by a straight edge in strict mathematical fashion and then went on to show that the previous approximations were justified.

In the beginning, light vibrations were regarded as elastic waves, similar to the transverse oscillations in solid bodies. No other concept was possible at that time. The medium, which was supposed to carry the vibrations through empty space, was called the ether. Ever since E. Torricelli and O. v. Guericke had produced fairly high vacua (sec Chapter II), there was no doubt that light, in contrast to sound, needed no material medium for its transmission. Of course, it was then difficult to explain why only transverse, but not also longitudinal waves, occur in the ether; bodies, in which only longitudinal but not transverse waves are possible, were known in the liquids and gases. No elasticity theory could account for the reverse case. Likewise, there was no complete mathematical solution for the problem of reflection and refraction. Then, in 1865 J. Clerk Maxwell (1831-1879), on the basis of his theory of electricity and magnetism (Chapter V), drew the mathematical conclusion of the possibility of electromagnetic waves, which travel with the velocity of light, and immediately he took light as an example. The electromagnetic theory accorded with experience better than the elasticity theory in so far as it permitted only transverse waves, and it relegated the difficulties of presenting the mechanical properties of the ether to the more general problem of a mechanical elucidation of electrodynamics as a whole. Furthermore, Maxwell's theory led to a simple relation, which fitted the empirical findings in many bodies, between the index of refraction and the dielectric constant, and it contained, as was shown in 1875 by Hendrick Antoon Lorentz (1853-1928), the complete theory of the Fresnel intensity formulas for reflection and refraction, which had been verified experimentally but which had not been explained by the theory of elasticity. Despite these advantages, it had to battle for three decades to secure acceptance, because the older theory, supported by the general mechanical conception of nature, was so firmly entrenched. After Heinrich Hertz discovered electromagnetic waves in 1888 and showed that they exhibit all the characteristics of light – refraction, reflection, interference, diffraction, polarization, and also travel with the velocity of light – victory gradually veered to the new theory. The long standing dispute as to whether the light vibrations occur in the plane of polarization, as postulated by Fresnel, or normal to it, as stated by Franz Neumann (1798-1895), was decided by the Lorentz theory of reflection and also by an experiment on stationary light waves made by Otto Heinrich Wiener (1862-1927) in which it was shown that the electrical field intensity vibrates perpendicularly to this plane, the magnetic strength within it. This uncoerced coalescence of the theories of light and electrodynamics, which hitherto had been completely independent,[1] is one, and perhaps the greatest, of those events to which the Introduction referred as proofs of the truth of

---

[1]EDITOR'S NOTE: A similar situation occurred in 1908 when Hermann Minkowski demonstrated that space and time are not independent but are aspects of an absolute entity – spacetime (Minkowski called it die *Welt*, i.e., the World).

physical knowledge.

Although Maxwells original theory gave a complete account of the propagation of light through empty space, it did not cover the optical properties of matter satisfactorily; in particular, it did not explain the dispersion of the refractive index. The molecular transformation into the 'electronic theory" which was due primarily to Joseph Larmor (1857-1942) and H. A. Lorentz, furnished the necessary supplement to the earlier theory. It accounted not only for dispersion, but also for a phenomenon discovered in 1871 by August Kundt (1839-1894), namely, the anomalous dispersion accompanying selective absorption, which it treated as a resonance phenomenon of atomic structures capable of oscillating. Tins electron theory had its greatest triumph when, by its aid, H. A. Lorentz elucidated the discovery (October, 1896) by Peter Zeeman (1865-1943) that spectral lines can be separated in magnetic fields. Magnetic rotation of the plane of polarization, which had been discovered as early as 1845 by Faraday, is intimately related to the Zeeman effect. There had now been developed a theory which in completeness was not inferior to that of mechanics and which took account of all the phenomena connected with the propagation of light. However, only three years later, it was found inadequate to deal with the facts of absorption and emission of light. These will be discussed in Chapters XIII and XIV.

5 ELECTRICITY AND MAGNETISM

The science of electricity and magnetism is much younger than mechanics and optics. Antiquity contributed nothing beyond the word magnet and some elementary observations concerning rubbed amber. Besides the compass, which can be traced back to at least the second century A.D. in China and which was brought to Europe in the thirteenth century, the Middle Ages added only the discovery that every part of a magnet again constitutes a whole magnet. It is perhaps worth noting the horror with which Christopher Columbus, on his voyage of discovery in 1492, observed the change from the easterly compass declination, which then prevailed in South Europe, into a westerly one. Even the first century and a half of modern physics provides the history of physics with very little in this field, despite the unquestioned services of William Gilbert (1540-1603) who coined the word electricity. He traced out the course of the lines of force by bringing a small compass needle near magnetized steel balls, thus demonstrating the complete analogy to the action of the earth on the compass. He put an end to all tales of great magnetic mountains at the North Pole or of a directing force coming from the lodestar. This condition was not altered much by the studies of Otto von Guericke, who noted the repulsion of like-charged particles, and constructed the first frictional electrical machine, and also discovered that iron filings can be magnetized merely by the action of the terrestrial magnetic field. In mechanics, optics, heat, and chemistry, there was available a fund of ancient pre scientific experience, which had a certain value to the budding planned investigations, whereas the orderly study of electricity and magnetism had to pass through the corresponding prehistoric' stage itself before it could lead to clear ideas. The investigators in the seventeenth and the first part of the eighteenth centuries were confronted by a confused collection of phenomena, such as frictional electricity, formation of sparks, and effect of atmospheric moisture, which they were unable to clarify because of the lack of fundamental electrostatic concepts.

Nevertheless, a number of important qualitative observations came from this period. In 1731, Stephen Gray (1670-1736) recognized the difference between conductors and insulators, and in 1759 Franz Ulrich Theodor Äpinus (1724-1802) went further by showing the existence of all stages of transition between them. They made the first observations of the influence exerted by charged bodies on insulated conductors. Both Ewald Georg von Kleist (born soon after 1700, died 1748) working in Kammin, Pomerania, and Pieter van Musschenbroek (1692-1761) in Leyden, Holland, were led by chance observation in 1745 to the Leyden jar, the original form of the electrical condenser, whose elucidation occupied not only Äpinus (Aepinus) but also Benjamin Franklin. (The latter was the author of the designations: positive and negative electricity.) In this connection, Johann Carl Wilcke (or Wilke) (1732-1796), in 1758 discovered the polarization of dielectrics, a typical instance of a premature and therefore quickly forgotten fact. Alessandro Count Volta (1745-1827) described in 1775 the electrophorus from which influence electrical machines were later developed. Great and deserved excitement was occasioned in 1752 when Franklin furnished the experimental proof of the electrical nature of thunder storms, a fact that had long been suspected.

The concept quantity of electricity seems to have been current since the seventeenth century, and without any real basis from the start was connected with the idea of the impossibility of creating or destroying electricity. The dispute as to whether there are two electrical fluids, which compensate each other in their effects, or only one, which is present in an electrically neutral body in a certain normal quantity, could not be decided. The present concept is dualistic, in so far as it ascribes different carriers to positive and negative charges, and unitary, in so far as it accepts the most elemental carrier of positive charges, namely, the atomic nucleus, as also being the fundamental constituent of matter.

During the eighteenth century, there was really only one discovery concerning magnetism, and it too was premature and consequently ineffective. In 1778 Anton Brugmans (1782-1789) discovered diamagnetism, when he found that bismuth is repelled by a magnet.

Electricity did not attain the rank of a science until the announcement of Coulomb's law: The force between two charges is directly proportional to the charge on each and inversely proportional to the square of the distance between them. This law had a peculiar history. It began with suspicions related to the Newtonian law of attraction. However, in 1767 Joseph Priestley (1733-1804) found compelling evidence of it in the discovery by him and others, Henry Cavendish, for instance, that the charge of a conductor resides entirely on its surface, while the interior remains completely free of all electrical influences. However, this fact received no attention. In 1785 Augustin de Coulomb (1736-1806) made measurements with a torsion balance; he determined the force between charged spheres, partly by means of the static deflection of the balance, and partly through the vibration period, about its rest point, of a sphere suspended from the balance and set in motion by the action of the fixed sphere. However, in a subsequent (1786) communication. Coulomb reported that a conductor also shields its interior – he was not aware of his predecessors – and he saw in this also an indication of the force law. This portion of his communication, however, was so completely forgotten, that the shielding action is now commonly linked with Faraday's name. In fact. Coulomb's contemporaries remembered only the more obvious measurements secured with the torsion balance, and the law which bears Coulomb's name was derived from these data.

When Coulomb stated that the force between two charges is proportional to the quantities of electricity, he did so purely in analogy to the Newtonian law. He could offer no proof for his assertion, because means of measuring charges were not available. The idea of defining this quantity directly from Coulombs law was contributed by Gauss.

Coulomb also extended his law to magnetism. In this case, however, the experiments were less convincing, because the accumulation of the magnetic fluida at the punctiform poles always remained somewhat in doubt, even though he had attempted to prove this by preparatory measurements. The valid portion of this extension finds its precise statement in the applicability of the Laplace differential equation to magnetism, as pointed out by George Green (1793-1841) in 1828. Coulombs proof, by means of the torsion balance, that the earths magnetic field exerts a moment of rotation on the compass needle proportional to the sine of the deflection from the meridian was important, because this finding constitutes the basis of the concept of magnetic moment.

The progress initiated by Coulombs law was shown in 1811 when Simeon Denis Poisson carried over to it the theory of potential (see Chapter III) which was first developed for gravitation. In fact the whole of electrostatics, in so far as dielectrics are not involved in the phenomena, can be covered by means of the Coulomb law or the equivalent Laplace-Poisson differential equation and the knowledge that the potential on conductors is constant. The further elaboration of the theory of potential was due, in addition to Green, to Karl Friedrich Gauss, who published a famous work in 1839. This theory had an effect far beyond its own sphere, because it became the model for many other fields of mathematical physics.

Gauss contributed not only the definition of the quantity of electricity from the Coulomb law, as was mentioned above, but he also provided the first absolute measurement of magnetic moment of steel magnets, and of the strength of the earths magnetic field. His mathematical theory of this field constitutes the direct and conclusive continuation of the work of W. Gilbert (page 51). He created with this the first rational electrical and magnetic system of units. In it, unit quantity of electricity is that quantity which, at a distance of one centimeter, repels an equal quantity with a force of one dyne.

The law of the conservation of electricity was first demonstrated in 1843 by Faraday. He brought a charged metal ball, suspended by a long silk thread, into an insulated ice pail, which was electrically connected with an electrometer. The resulting deflection of the electrometer is a measure of the charge. He then showed that this deflection is independent of any other contents of the "ice pail and also of what happens to the charge there. All or part of it can be transferred to other conductors; there is no effect. Only when additional charges are brought into the pail is there a change in the deflection in that the instrument now indicates the algebraic sum of charges that have been introduced. Other demonstrations of the law of the conservation of energy were being made around this time, and even though this experiment was not less important than the others, it did not receive the same recognition because the doctrine of the indestructibility of electrical fluids had already firmly established itself and hence needed no champion to do battle in its behalf.

The second and perhaps more fruitful advance was made by the science of electricity when Alessandro Count Volta transformed into a physical discovery the observations on frog legs reported by the physician Luigi Galvani (1737-1798), which had aroused much interest and had stimulated many successors. Seldom has a discovery posed so many difficulties to the human understanding as this one, but, in return it opened the way into utterly undreamed-of territory.

Galvanis first chance observations on the contraction of frog muscles connected with a metal loop in the neighborhood of electrical spark discharges at the approach of a thunderstorm were actually the earliest indication of electrical oscillation; the frog leg acted as a detector. The physicists did not really accomplish anything with this observation until 100 years later. In 1792, however, conscientious experimentation and good fortune brought Galvani to the point at which he was able to make the muscle contract by simply applying to it a loop consisting of two different metals. This was the first galvanic element; the frog muscle was both its electrolyte and current indicator. Galvani himself, however, remained ignorant of these facts; he supposed, and perhaps not entirely erroneously, that these were manifestations of animal electricity, which had been known for a long time in the case of electric eels and other fishes.

Volta also accepted this interpretation in 1792. However, long series of experiments brought him more and more to the conviction that the biological object, frog leg or human tongue, was only of secondary importance. In 1796 he eliminated it entirely and stated that an essential condition for the circulation of electricity in a conducting circuit was that the latter consist of two (or more) conductors of the first class and one of the second class. He originated these ideas, as well as the concept of the stationary electric current. On the basis of the new knowledge, he constructed in 1800 the voltaic pile, the prototype of the galvanic batteries, which in the succeeding years and decades sprang up like mushrooms from the ground. Voltas fundamental experiment, which was designed to demonstrate the charging of two metals on contact, became famous. It has not stood the test of modern critical examination (Emil Warburg); invariably there is a layer of moisture between the metal plates and what is actually observed is the terminal voltage of an open galvanic cell. However, Voltas observation was correct and fundamental, because an electrical equilibrium, which excludes all current, is set up momentarily in a purely metallic circuit, no matter how many different metals are included in it. The fact that temperature differences also produce a current in such circuits (thermoelectricity) was first observed (1821) by Thomas Johann Seebeck (1770-1831).

Electrolytic decomposition, which is now regarded as the

cause of the production of galvanic current, was described in 1797, i.e., before the voltaic pile, by Alexander von Humboldt (1769-1859), whose fame ordinarily is ascribed to his achievements in the descriptive natural sciences. His discovery was made with a cell consisting of a zinc and a silver electrode with a layer of water between. This finding was further utilized in 1799 by the gifted but visionary Johann Wilhelm Ritter who, for instance, separated copper electrolytically from a solution of cupric sulfate. He also recognized the identity of static and galvanic electricity in that he employed the discharge of a Leyden jar for electrolysis. He likewise was the first to bring forward the idea that chemical reaction in the galvanic cell is the cause of the production of the current. In 1800 Humphry Davy began his famous electrolysis researches, which, for example, led him to the discovery and separation of the alkali metals in 1807. His quantitative determinations of the amounts of the decomposition products opened a new field of investigation. This yielded: in 1834, the Faraday law of electrochemical equivalence; in 1853, the studies on the migration of ions by Johann Wilhelm Hittorf (1824-1914); also, in 1875, the recognition of the independence of ionic mobilities by Friedrich Kohlrausch (1840-1910); in 1887, the theory of electrolytic dissociation by Svante Arrhenius (1859-1927). The theory of electromotive forces of Walter Nernst brought this glorious series to a worthy close in 1889, and the theory of the galvanic production of current was thus completed. Of course, the idea that a sodium ion, for example, can move freely in a water solution without reacting chemically with its surroundings aroused considerable heated opposition at first; many refused to accept the distinction between the neutral atom and the ion. However, the Arrhenius theory received so many substantiations that the opposition was gradually silenced.

Voltas discovery opened still other lines of development. The galvanic cells produced electric currents of strengths and duration, quite different from those that had been obtained from the discharge of condensers and similar devices. For example, about 1811 Davy constructed the carbon arc with a battery of 2000 elements, and it served as a source of electric light until Thomas A. Edison (1847-1931) invented the incandescent bulb around 1880. Batteries also made the magnetic action of currents accessible to study. Surmises concerning the forces between the electrical and magnetic fluids were rather common at the beginning of the nineteenth century and had occasioned, for instance, the search for mutual actions between magnetic poles and open voltaic piles. Independent of such wrong turnings, and quite by chance, Hans Christian Oersted (1777-1851) in 1820 came upon the deflection of the compass needle by the electric current, and thereupon also discovered the corresponding directive force of a magnet on a rotatable electric circuit. Many others, especially French physicists, now entered the newly opened field, and the foundations of electromagnetism were laid in the short time up to 1822. First came the observation by Dominique Francois Arago and Joseph Louis Gay-Lussac (1778-1850) that a piece of iron is magnetized by a current flowing in a wire looped around it; this was the first electromagnet. Later, in this same year (1822), André Marie Ampère (1775-1836), who took no part in physics either before or after this period, set up his familiar swimmer rule for indicating the direction of the magnetic field of an electrical circuit and discovered that parallel currents passing in the same direction attract each other, and repel each other if their directions are contrary. He showed that a solenoid acts like a bar magnet. Jean Baptiste Biot and Feliz Savart (1791-1841) concurrently formulated from the experimental findings the law bearing their names, which deals with the magnetic action of a single line clement of a linear current. In 1822 Faraday caused the movable part of a circuit to rotate by the action of permanent magnets and imparted rotatory motion to magnets through the action of currents. Thereupon, Ampère in 1822 demonstrated the rotational effects of two circuits and used this as the starting point for his fundamental law of electrodynamics," a term that appears for the first time in his writings. However, more than a century elapsed before his explanation of magnetism became especially important; he abandoned the hypothesis of magnetic fluids, and (1821-22) ascribed magnetism to the action of his hypothetically assumed molecular currents.

These magnetic effects of currents now provided a means of measuring current strength. This was employed in 1826 by Georg Simon Ohm when, clearly separating the concepts of electromotive force, potential gradients, and current strength, he derived the (Ohms) law of the proportionality between current strength and difference of potential, in which the proportionality factor represents the resistance of the conductor. He proved that in the case of a wire the resistance is proportional to its length and inversely proportional to its cross section and thus created the basis of the concept of the specific conductivity of materials. The latter, however, is one of the three constants which characterize the total behavior of every substance toward electricity and magnetism. In 1847 G. R. Kirchhoff solved the problem of branched circuits and set up the rules that now bear his name.

An application of these effects came when the telegraph was invented. The form devised in 1833 by Gauss and Wilhelm Weber (1804-1891) differed from its predecessors in that the current returned through the earth.

The development of electromagnetism halted after 1822, even though at first only half of this group of phenomena had been recognized. In 1831 Faraday wound two coils of wire around an iron ring and with this arrangement found that currents exert a back action which corresponds to their magnetic action. When he sent a current through the first coil, a pulse of current appeared in the second at the instant the circuit was closed, and again when it was opened, but in the reverse direction. In this way he discovered induction, and he clarified its various kinds in the years that followed. Faradays somewhat scattered statements regarding the direction of the currents induced by movements were summarized in 1833 in the well-known (Lenz) law by Friedrich Emil Lenz

(1804-1865). Induction machines soon followed, so that currents could be produced independently of galvanic batteries. However, their large-scale development did not come until after 1867, when Werner von Siemens (1816-1892) substituted electromagnets, fed by the produced current itself, for the steel magnets previously used. This constitutes the dynamo-electrical principle.

Electrodynamics now made possible a second system of electrical units; it is independent of Coulombs law. For example, the unit of current strength can be defined as the current which flows in two long parallel wires, one centimeter apart, when they exert on each other a force of two dynes per unit length. Since unit current strength must furnish unit quantity of electricity to a condenser, in unit time, an electromagnetic unit quantity of electricity has likewise been obtained. This necessarily raises the question as to the relation to the electrostatic unit as defined by the Coulomb law. An examination of the formulas shows that this relation has the dimension of a velocity. Wilhelm Weber determined its value in 1852 and the succeeding years, with the astounding result that it is equal to the velocity of light: 3×10^{10} cm/sec. As it was of fundamental importance to the electromagnetic theory of light, James Clerk Maxwell redetermined it in 1868-69 with greater accuracy. Subsequently, the comparison was perfected to such a degree that it is now included among the precision measurements of the velocity of light.

An international congress held at Paris in 1881 set up the electrical units (ampere, ohm, volt, etc.) now used in technical practice. These are based on the electromagnetic system of units. The prospects of any considerable technical development were so slight at that time that these authorities did not adopt the electromagnetic unit of current itself, because it seemed impractically large. They accordingly defined the ampere as one-tenth of this value.

All the measurements cited hitherto dealt with currents passing along metallic conductors. In 1872 Henry Rowland (1848-1901) showed that the convection currents of static charges exert the same effect on bodies in motion.

The discoveries of electrodynamics confronted theory with problems, which, in contrast to all those that had arisen previously, could no longer be solved solely by means of the central forces depending on the distances between mass points. Ampère and Franz Ernst Neumann, and, above all, W. Weber, attacked these problems. Weber's fundamental law (1846) assumed that the force between two charges depended not alone on the distance between them but also on the velocity and acceleration and that currents are moving charges. It covered electrostatic as well as electrodynamic forces including induction in closed circuits, in other words, everything that was known about electricity at the time. Consequently, this law played an important role in science until about 1890. However, all these theories contained the defect of assuming action at a distance and as soon as the finite speed of propagation of electrical actions was recognized, the ground was cut from under them. Today they merely illustrate the difficulty that beset the path of progress in this field, and show the extent of the great changes in the whole physical viewpoint that have transpired since their day.

Michael Faraday was the leader in acquiring the correct understanding of electrical and magnetic phenomena. In 1837 he discovered the influence of the dielectric on electrostatic processes and in 1846 and the following years the general distribution of the diamagnetic properties over all material to which, in contrast, paramagnetism appears as an exception. On this basis, he evolved the idea that electric and magnetic actions do not pass from body to body without a medium, but are transmitted through the dielectric which lies between and which accordingly becomes the seat of the electrical or magnetic field." This latter thought also came from Faraday. As his experiments progressed, this idea developed. "The method which Faraday employed in his researches consisted in a constant appeal to experiment as a means of testing the truth of his ideas, and a constant cultivation of ideas under the direct influence of experiment. In his published researches we find these ideas expressed in language which is all the better fitted for a nascent science, because it is somewhat alien from the style of physicists who have been accustomed to established mathematical forms of thought." This judgment is by J. Clerk Maxwell[1] who continues: It was perhaps for the advantage of science that Faraday, though thoroughly conscious of the fundamental forms of space, time, and force, was not a professed mathematician. He was not tempted to enter into the many interesting researches in pure mathematics which his discoveries would have suggested if they had been exhibited in a mathematical form, and he did not feel called upon either to force his results into a shape acceptable to the mathematical taste of the time, or to express them in a form which mathematicians might attack. He was thus left at leisure to do his proper work, to coordinate his ideas with his facts, and to express them in natural, untechnical language." Concerning his own researches, Maxwell then adds: It is mainly with the hope of making these [Faradays] ideas the basis of a mathematical method that I have undertaken this treatise."

It was in this sense that Maxwell, in a first paper of 1855-56, provided the appropriate mathematics for the Faraday idea of lines of force. Particularly through his analysis of the course of the magnetic lines of force in the vicinity of an electric current, he arrived at the now familiar vector differential equation, according to which every current path produces a vortex line of the magnetic field, although with limitation to stationary fields. The achievement that was most particularly Maxwells own, the step that was decisive for everything which came later, was first contained in a paper of 1862.[2] He added to the conducting current the displacement current, which occurs in every dielectric with varying electrical field strengths, and only in conjunction with this current, gives the total current which invariably is a closed whole. Actu-

---

[1] J. Clerk Maxwell, *Treatise on Electricity and Magnetism*, Vol. 2, Oxford, 1873, pp. 162, 163.
[2] *Philosophical Magazine* [4], 23, 12 (1862). (See equation 112.)

ally, Maxwell came upon this through a hypothetical quasi-mechanical model. Nobody would regard this derivation as compelling truth, and it was not included in Maxwells comprehensive textbook that was published in 1873. However, it is interesting to note that it was only by this roundabout way that he arrived at the decisive step. The electromagnetic theory of light (Chapter IV), i.e., the recognition that there are electromagnetic waves possessing the velocity of light, then became no more than a necessary conclusion; Maxwell drew it in 1865.

The transmission of force through the electromagnetic field was ascribed by Maxwell to the stresses which bear his name. Entirely analogous to the elastic strains, analyzed by Cauchy (Chapter II), they differ from these only in that they are not associated with deformation of matter, rather, occasioned entirely by the field, they may reside entirely outside of all matter, even in empty space. According to this viewpoint, in a purely electrical or purely magnetic field, there is a stress along every line of force, and an equally strong pressure perpendicular to it. Only the Maxwellian stresses bring the ideas of close action to completion.

The bases of the present-day theory of electricity were thus laid down completely. Of course, it was not until 1890 that Heinrich Hertz put the Faraday induction law into the differential equation form in which it appears as a counterpart of the differential relation given by Maxwell. As a result, the system of Maxwell equations, in which modern physicists, along with Hertz, see the essence of the Maxwellian theory, was given that absolutely esthetically beautiful symmetrical form, which in view of its comprehensive physical content seems almost to have the character of a revelation. And yet this was only a matter of form. New physical knowledge was not added to it until 1884, when Poyntings theory of energy flowing was put forth (Chapter V), and by the demonstration, around 1900, by H. A. Lorentz and Henri Poincaré (1854-1912) that an electromagnetic impulse is associated with the electromagnetic energy current. However, this represents only a slight supplement to but no fundamental alteration of the basic theory.

Despite its inner completeness and the agreement with all experiment, Maxwells theory was only gradually accepted by the physicists. Its ideas were too unconventional; even men of the caliber of Helmholtz and Boltzmann had to strive for years to secure an understanding of it. In 1879, the Berlin Academy offered as topic for prize competition the experimental proof of an effect of dielectrics on magnetic induction. Heinrich Hertz solved this problem in 1887 by means of rapid oscillations. Another result of such considerations was the experiment by W. C. Röntgen in 1888 to determine whether the motion of an electrically polarized dielectric has the magnetic effects of a current, as would correspond to Faradays idea. The effect which he definitely demonstrated is called the Röntgen current. All doubts were conclusively removed by Hertz through his discovery in 1888 of electrical waves. He directly determined the velocity of propagation from the frequency and wavelength and found that it equals the velocity of light.

The previous history of this discovery goes back to the Helmholtz essay The Conservation of Energy (1847, Chapter VI). Various studies of the discharge of , especially the independence of the heat produced in the discharging wire of all special characteristics of the wire, led Helmholtz to conclude that the discharge is oscillatory in character. Likewise, in connection with the energy principle, William Thomson (Lord Kelvin) in 1853 gave the mathematical theory for it in a form to which practically nothing has needed to be added. Berend Wilhelm Feddersen (1832-1918), from 1858 to 1862, examined these oscillations in the image of the discharge spark in a rotating mirror. Friedrich Wilhelm von Bezold (1837-1907) definitely observed oscillations in wires with one free end and in wire circuits containing a spark gap. But it was only in the hands of Hertz that such wire circuits became the means of studying waves in free air, for revealing their polarization, reflection, and refraction, as well as their interferences, which then made it possible to measure the wavelength and so to determine the velocity of propagation.

The waves with which Hertz experimented were strongly damped. If his experiments can be repeated today with undamped waves, i.e., with greater nicety, it is because of technical advances. However, technology had to tread a toilsome path before it learned in 1913 and later how to produce, by means of vacuum tube transmitters based on feed back (Chapter I), the undamped waves needed for wireless telegraphy and similar purposes.

Just as a period of mathematical development of mechanics followed Newton, a similar era of mathematical elaboration of the Maxwellian theory now set in. The vector potential had been introduced even earlier to represent the magnetic eddy fields in the vicinity of stationary currents. In opposition to this and the scalar potential of electrostatics Alfred Marie Liénard in 1898 and Emil Wiechert (18G1-1928) in 1900 proposed the retarded potentials, in which the finite velocity of propagation of magnetic actions finds its most striking expression. The available space would by no means permit the enumeration of all the investigators who elucidated mathematically the scientific and soon also the technically important cases of electrical alternating fields. In its modern form the Maxwellian theory is an inspiring masterpiece, fully the peer of mechanics.

Thus, at the beginning of the twentieth century, the theory of electricity and magnetism seemed to be fairly complete, especially since atomistics had shortly before brought order and clarity into the confusion attending the phenomena of discharge through attenuated gases (Chapter X). And yet, a new and unexpected phenomenon appeared in its own most particular province, in the conduction of current. It had been known since 1835, from measurements by Heinrich Friedrich Emil Lentz, that the resistance of metals decreases when they are cooled, Heike Kammerlingh-Onnes (1853-1926) followed the decrease to below $10°$ absolute, when the attainment of such low temperatures became possible through his liquefaction of helium in 1908. Metals, e.g., gold, silver, and copper, were found to have a limiting value below which the resistance does not fall. However, in 1911 he observed, first with mercury and later with lead, tin, and several other metals, a sudden disappearance of all resistance as soon as the temperature fell below a transition point whose value is characteristic of the substance. Supraconductivity sets in; the current then flows without any potential gradient, and persists with undiminished strength and without any electromotive force as a permanent current for days in a supraconducting ring. This was demonstrated in 1914 by Kammerlingh-Onnes, He eventually also found that, without change in temperature, it is possible to annul supraconductivity by means of a magnetic field; Ohm's law then holds as usual. The field strength, which the supra-conductor just withstands, the threshold value," increases the lower the temperature is brought below the transition point. With pure metals it amounts to several hundred gauss.

Later investigators added a few more pure metals to the list of supraconductors and also a series of alloys and chemical compounds. W. J. de Haas and his associates observed further that the threshold value of a supraconducting wire seemed to depend on the direction of the magnetic field with respect to its axis. Max von Laue provided the explanation of this in 1932: a homogeneous magnetic field is deformed when a supraconductor is brought into it, because the lines of force avoid the conductor, in accordance with the Maxwellian theory, as Gabriel Lippmann (1845-1921) had previously concluded, The crowding together of the lines of force, however, brings about a strengthening of the field at certain points of the surface; the supraconductivity fails as soon as the threshold value is attained at even a single point. Subsequently, de Haas and his collaborators quantitatively verified this explanation with supraconductors of various forms.

A supraconductor is not a conductor in the usual sense of the Maxwellian theory that is merely distinguished from others by possessing an infinitely great conductivity. If this were so, a magnetic field that is entered above the threshold value must continue to exist inside the conductor when the temperature is lowered below this point. However, in 1933 W. Meissner and R. Ochsenfeld made measurements which showed that the field is forced out. It makes no difference whether the temperature is brought below the threshold value before activating the magnet, or vice versa. This Meissner effect requires a supplementation of the Maxwellian theory on a completely new foundation.

The relation of the electromagnetic field to its charges was subjected to remarkable fluctuations in the views of the physicist. Just as gravitation seemed to Newton and his successors to be the causal consequence of masses, every physicist at first conceived electrical forces as due to charges. Then the field concept came to the front with Faraday and Maxwell, and the charges were demoted to a kind of singular areas of the field. But the relation was again reversed when the rise of the electronic theory of optics and electrical discharge put the atomic carriers of the electrical elementary quanta in the forefront of interest. Neither of these viewpoints seems to fit the facts. Charges and field are so closely associated with each other that one cannot exist without the other. Precisely for this reason, science can just as well take the charges as the criterion for the knowledge of the field as to draw a conclusion from the course of the electrical lines of force. These are logical conclusions; they have nothing to do with the real relation of cause and effect. The same applies, of course, to the gravitational field and its masses.

The relations between the theory of electricity and mechanics are also of a special kind. As was mentioned. Maxwell around 1862 tried to construct for himself a mechanical picture of the magnetic field. Later, during the progressive acceptance of his theory, many sought in a more rational way to base it on a mechanics of the ether. To a certain degree, it is possible to subordinate the theory of linear, closed (quasi-stationary) currents to the theory of cycles, derived from mechanics, and developed especially by Helmholtz. This is little more than a mathematical analogy between different varieties of physical events. After all, it is indicative of the infiltration of electrodynamic views into wide circles that the present-day engineer frequently explains the mode of action of mechanical machines through a corresponding electrical connection. However, around 1900, it was gradually perceived that a general reduction of electrodynamics to mechanics is impossible.

Since 1880 the reverse idea has gradually taken shape, i.e., to refer mechanics back to electrodynamics. The fact that a moving charged body carries its electromagnetic field with it, and that an impulse resides in this, certainly was close to the idea of an electromagnetically inert mass. In fact many workers tried to conceive of every mass as electromagnetic mass. This found, for instance, its mathematical downfall (1902) in the theory of Max Abrahams (1875-1922), which deals with the impulse of the moving electron treated as a charged sphere. The mass proved to be dependent on the velocity, and the Abrahams formula covering this was in competition with the relativity formula for a long time (Chapter II).

Physics has discarded this idea also, since experiment has finally definitely decided in favor of the relativistic formula. In addition, the Abrahams theory produced a factor for the proportionality between energy and the mass at rest[3] which differs from that appearing in the Einstein law of the inertia of energy. The latter has been fully confirmed through nuclear physics (Chapter XI). However, the Abrahams investigations had a permanent effect as preparation for relativistic dynamics.

Even though relativistic dynamics is completely independent of every concept concerning the nature of forces, i.e., also independent of electrodynamics, nevertheless the latter

---

[3]EDITOR'S NOTE: Here von Laue has in mind the famous 4/3 factor. However, even without employing relativity, if the calculations are properly done by taking into account the correct (anisotropic) volume element, this factor does not appear; see V. Petkov, *Relativity and the Nature of Spacetime*, 2nd ed. (Springer, Heidelberg 2009), pp. 247-249.

played a decisive role in the discovery of this dynamics. The findings which led to the Newtonian dynamics could never have sufficed to produce the Einstein relativity theory; they were not accurate enough. When electrodynamics compelled the relativity principle that is connected with the Lorentz transformation, it also compelled the change from Newtonian to relativistic dynamics. Therefore, in this purely historical sense, modern dynamics is based on electrodynamics.

6 THE REFERENCE SYSTEM OF PHYSICS

The problem, suggested in the chapter heading, can be traced back into Greek antiquity. It had two epochs: the geometrical, which lasted into the seventeenth century; and the dynamic, which, after the triumph (around 1800) of the wave theory of light, spread out until it embraced all of physics.

In the geometrical period, the question of the reference system was directly linked with the nature of the location and motion of a body. From the very beginning, it was apparent that both concepts were without meaning in the absence of something to which they could be referred. Thus Aristotle, and with him all of the scholastics, located a body on a material structure which embraced it; moot points being whether this structure had to be in immediate contact with the body or whether finite distances between them were allowable. Disputes arose over such questions as to whether a ship anchored in a stream moves when the wind blows, since the water and air in the ships immediate vicinity constantly change, or whether the vessel is at rest, since it exhibits no movement that can be seen from the bank. Such questions could not be decided under the prevailing conditions. From the standpoint of physics, it was more important when Claudius Ptolemy, who lived at Alexandria in the second century A.D., declared that the sphere of the fixed stars, the outermost of a series of spheres surrounding the earth, had no location whatsoever, there was nothing at all beyond it, not even space, which should have enveloped it. It is a quite remarkable feature of the Ptolemaic system that it insisted that this sphere nevertheless possessed motion, i.e., a daily rotation around the earth.

Nikolaus Copernicus noted this inconsistency when he founded the astronomical system which bears his name. When he declared that the sphere of the fixed stars remains at rest, while on the contrary the earth rotates daily about its axis, this was, at first, no more than an improvement of the consequence of the idea that had come down to him. He also firmly retained the notion of the existence of utter void beyond this sphere, an idea which the moderns find exceedingly difficult to comprehend. The first to free himself of this notion was the powerful attacker of Aristotle, Giordano Bruno, who in 1600 was burned at the stake at Rome as a penalty for teaching the related doctrine of the unlimited multiplicity of worlds, his defense of Copernicus, and similar heresies. Not even a man of Keplers stature was able to adopt this bold and yet so unavoidable addition to the Copernican system. In the further establishment of the heliocentric system, Copernicus allowed himself to be guided by the, so to speak, teleological viewpoint of the simplicity of Nature; even the Greeks were familiar with the phrase Nature does nothing superfluous or in vain. He found it simpler to explain the occasional retrograde motion of the planets in the heavenly vault as due to the movement of the earth itself, i.e., of *one* body around the sun, rather than through the motions compounded of several circular movements of *all* the planets, as postulated by Ptolemy. Though this thought is congenial to the modern scientist, the state of physics at that time was such that nobody could provide a causal basis for it. Though appealing more to sentiment than to reason, this demand for simplicity was incapable – as is entirely understandable – of convincing many of his contemporaries and those who came later, principally because the idea that all humanity was being whirled around in a circle without being conscious of the fact was not exactly simple and, in addition, at that time without foundation in physics. Hence, by no means need it to have been mere backwardness or anxiety which impelled Osiander, the Nuremberg scholar, who supervised on the very spot the publication of the great Copernican treatise De revolutionibus[1] to state in the Foreword, which he supplied to the work, that the Copernican system was being presented as a pure hypothesis,"' which, although it was in accord with the observations and was therefore justified to that extent, was by no means necessarily true because of this agreement. The heat of the controversies which this system quickly engendered likewise was due, in no small measure, to the total lack of physical-causal reasons, either for or against it, during the entire geometrical epoch. An additional factor, of course, was the opposition of the representatives of the church, Protestant as well as Catholic. They, Dr. Martin Luther, for instance, denied the movement of the earth as being contradictory to the Bible. Actually they were only repeating the indictments, that had been leveled in the third century b.c. against the first champion of such a system; Aristarchus of Samos had likewise been labeled scoffer at religion. Experience has revealed a very singular fact, namely, that – from Aristarchus to Einstein – nothing physical embitters wide circles of the public so much as an attack on firmly entrenched concepts of space and time. Similarly, Galilei's condemnation was not connected with the geometrical arguments which Copernicus had to cite for his system nor with Galileis astronomical discoveries, which he utilized in its support, but rather he was condemned because of the Dialogue on the Two Chief Systems of the World, which refutes the dynamic reasons against the movement of the earth in detail and sometimes caustically.

Coernicus himself was not touched by the disputes concerning his system. He had held back his book, begun in 1507, until 1543, and at most saw parts of it in print on his deathbed. To be sure, much earlier, probably in 1514,

---

[1]The last two words in the title "De revolutionibus orbium coelestium" were added by the publisher himself.

he had sent a friend a sort of preliminary announcement in manuscript form. This later became known as the *Commentariolus*; after being forgotten for many years, it turned up in the Vienna Library about 1880. The following rules are taken from it;

1. There is not merely one central point for all heavenly circles or spheres.

2. The center of the earth is not the center of the world, but only of that of heaviness and the orbit of the moon.

3. All orbits surround the sun, as though it lies in the middle of all, and consequently the center of the world lies near the sun.

Obviously to meet the objection that the appearance of the heaven of the fixed stars must change in the course of a year because of the terrestrial motion, there then follows:

4. The relation of the distance sun-earth to the height of the fixed star heaven is smaller than that of the earths radius to its distance from the sun, so that this is imperceptible in comparison with the height of the fixed star heaven.

5. All that is visible of motion in the fixed star heaven is really not within itself, but is seen from the earth. Hence, the earth, together with its contiguous elements, rotates once each day about its invariable poles. Meanwhile, the fixed star heaven remains immovable as the outermost heaven.

6. All that is visible of motion on the sun does not arise through itself, but because of the earth and our circular orbit, with which we, like every other planet, revolve around the sun. And thus the earth is carried along by several motions.

7. The seeming retreat and advance of the planets is not within them, but are seen from the earth; its motion therefore is sufficient to account for so many varied phenomena in the heavens.

This now appears to be clear and simple, and yet the conception of such a system still gave Copernicus trouble. How otherwise can it be explained that, in addition to the daily and yearly rotation of the earth, he later also ascribed to it a third motion? This was supposed to account for the fact that the earths axis changes its position relative to the sun during the course of the year, i.e., it was to explain the change of the seasons. It would have sufficed, however, if the earth maintained its position relative to the heaven of the fixed stars. This error is somewhat reminiscent of the controversy current in Newtons time as to whether the moon has a rotation of its own, since it always presents the same side to the earth. Those who denied this rotation simply could not free themselves completely from their ingrained geocentric viewpoint. A certain prejudice in favor of the sun certainly played a similar part in the case of Copernicus.

Be that as it may, we owe to Copernicus the reference system, which rests in the center of gravity of our solar system, with its axial directions oriented toward the heaven of the fixed stars. Physics refers all locations and all motions to this system, if nothing is stated to the contrary. The three coordinates, by which the point is defined in terms of Cartesian analytical geometry, in physics are referred to this system, if they are not expressly defined in some other manner. Neither Kepler's laws nor the gravitational theory would have been discovered without Copernicus. It is conceded that its foundation remained incomplete; newborn sciences cannot be established in their entirety from the very start. The greatness of their creators is revealed precisely in the fact that they, nonetheless, intuitively hit upon the right thing.

Johannes Kepler contributed to the consolidation of the Copernican system to the extent that his three laws of planetary motion, and the more exact computation of observations which they made possible, had been nurtured wholly in its soil; furthermore, they could scarcely have fitted into the Ptolemaic system. However, Kepler had little inclination to establish the Copernican system more firmly; his arguments in its favor go not much beyond pointing out its simplicity and beauty, features which Copernicus himself had thrown into the scales. Although Kepler received the reports of the astronomical discoveries of Galilei with delight, he seemingly had little understanding of his contemporarys contributions to dynamics, despite their importance in deciding between Copernicus and Ptolemy, and in the face of the fact that dynamical arguments against the Copernican system had quickly been raised by its opponents. They stated, for instance, that the daily rotation of the earth around the sun would surely cause everything that was not securely fastened to the earth to be hurled into space, and that such articles because of the motion around the sun would be brushed off by the earth and so trail behind. Objections of this type could only be successfully met by the new dynamics. The decision came partly from Galilei, and the matter was finally settled by Newton.

In his time, Galilei was the most successful, most popular, and most hated champion of the Copernican system. His contribution lay first of all in his astronomical discoveries which were made possible by the use of the new tool, the telescope. In 1610 he showed that Jupiter and its satellites are a Copernican system in miniature. In 1611 he definitely proved by means of the phases of Venus that this planet follows an (approximately) circular orbit around the sun and that it, like the earth and the moon, does not emit any light of its own, but merely reflects the light falling on it from the sun. The ashen-gray light sent out by the part of the moon not illuminated by the sun was proof to him that the earth, if viewed from without, would appear bright like the other planets. Secondly, he had the clear insight to comprehend that the laws of motion should not be related to an earth-bound reference system, but to the Copernican system. Although when discussing the falling of bodies, he usually spoke of motion toward the earth, he nevertheless added that this is only approximately permissible, that in the strictest sense, a freely falling body deviates from the vertical because of the rotation of the earth. This approximation, whose limitation is fully recognized, is still employed by physicists in the discussions of most of their experiments. Galilei never tired of using his newly acquired knowledge in refuting the dynamic objections against Copernicus. However, Newton's work made it perfectly evident that the planetary motions can be understood dynamically – but then also completely – only when the Copernican reference system is taken as the basis.

The problem of the reference system had thus been decided practically, but not fundamentally. By what physical authority is the Copernican system of reference given precedence over all others, an earth-bound one, for instance? Newton, who was thoroughly aware of the gravity of this question, had recourse to the assumption that there is an absolute space, just as there is an absolute time, and that it precisely defined the correct reference system. Most people, however, are inclined to agree with Ludwig Lange (1863-1936), who found both concepts hard to grasp and rather ghostly; nonetheless, they still haunt some minds today. As a matter of fact, the debatableness of this idea gave rise to thought by all the great philosophers from Newtons time on – Leibniz and Kant, for instance. But the liberating word did not come until 1886 when Langes The historical development of the concept of motion appeared. He wrote: Physics defines its reference system according to the purpose which it is to fulfill, hence, from the same viewpoint which is also the basis of measuring time. Lange summarized the result of his deliberations in two definitions and two theorems.[2]

Definition I: An inertial system means every coordinate system of such nature that with reference to it the paths of three mass points (which, however, may not lie on a straight line) projected from the same point in space and then left to themselves arc all straight lines.

Theorem I: With reference to an inertial system the path of every fourth point left to itself is also a straight line.

Definition II. Inertial time scale means every time scale with respect to which a mass point left to itself traverses equal distances in equal times in its inertial path.

Theorem II. With respect to an inertial scale, every other mass point also traverses equal distances of its inertial path in equal times.

---

[2]The quotations are from a section of his *Philosophischen Studien*.

The definitions are human convention, but the theorems are empirical propositions, and only through them do the definitions acquire value for physics. The truth of the Copernican system lies in the empirical validity of the theorems.

Of course, it was not possible to conclude from the observation of force-free mass points that the Copernican system is an inertial system. However, a perfectly valid proof of this resides in the agreement with experiment that is obtained when planetary orbits are computed on the basis of mechanics which contains the law of inertia. It was pointed out in Chapter I that, in order to attain nicety in this agreement, certain slight corrections must be applied to the ordinary time scale, so as to transform the latter into the inertial scale.

The work of Ludwig Lange thus finally brought a certain degree of conclusion to the development, which was begun three and a half centuries previously by Copernicus. The fact that the other natural laws, optical, electrical, etc., can be simply formulated in the same system, is, of course, a further purely empirical fact.

Many other conceivable reference systems are excluded by the above requirement of serving the purpose; for example, every system that rotates with constant velocity counter to the astronomical system. A body at rest in such a system – even Newton called attention to this possible case – would apparently experience a centrifugal force, of which the equations of motion contain nothing, and which actually is nothing more than another expression for the tendency toward linear motion in relation to an inertial system. In the case of the earth-bound reference system, which therefore rotates with the earth, this is manifested, for instance, by the flattening at its poles. The rotation of the plane of the swing of the so-called Foucault pendulum, which was first demonstrated in public in 1851, gives further incontrovertible evidence of the earth's rotation. In other words, this again proves that the fixed-on-earth reference system has no justification, as does the fact that the gyroscopic compass takes a north-south position. A. A. Michelson furnished another proof in 1925 by means of an optical interference experiment.

Still other systems, however, can be derived from one inertial system according to dynamics. All reference systems have a standing equal to that of any other system toward which they possess a constant translation velocity. Newton was well aware of this. In fact, Galilei, when defending the Copernican theory against popular mechanical objections, emphasized that no mechanical experiment will reveal the motion of a moving ship to a passenger inclosed within the vessel. The coordinates of a mass point in one system can be calculated by means of a simple formula from the coordinates with reference to the other. Time also enters in this process, of course, corresponding to the relative motion of both systems. It remains untransformed, and in this sense is "absolute." The velocities of the mass points are different in two such systems, but not their accelerations; consequently, the law of motion holds for both in exactly the same form. If there were a single inertial system, it could be put down as the absolute reference system, and motion with respect to it could be termed absolute" motion. Since this is not so, one speaks of the relativity principle of the Newtonian mechanics. The meaning of this principle is that mechanical experiments cannot reveal the superiority of one inertial system over another inertial system.

For a long time it was necessary to conjecture that perhaps other experiments and observations might permit this, namely, all those in which physical effect are propagated with finite velocities, since the velocities are different in different inertial systems. This idea became especially important for optics. Every theory which required an ether as carrier of the light – and eventually the electronic theory came to it (Chapters IV and V) – must, of course, regard the reference system in which it is at rest as superior to all others. It thereby defines an absolute reference system. That this is an inertial system in the sense of mechanics was always tacitly assumed.

Actually, the aberration of the stars, discovered in 1728 by J. Bradley (Chapter IV), could be simply explained on the basis that in the astronomical reference system the light travels in all directions with the same velocity which had been determined by Olaf Römer, and that its relative velocity with respect to the earth can be obtained from this through vectorial subtraction of the velocity of the earth. The fact, which seemed paradoxical at first, that filling the telescope with water, i.e., changing the speed of light within the tube, would have no effect on the aberration was predicted in 1818 by Fresnel and verified by an actual trial in 1871 by George Biddell Airy (1801-1892). Fresnels theory is no longer of interest, but it led to the correct result, verified in 1851 by A. Fizeau, for the velocity of propagation of light in moving bodies. If the ray and the motion of the body have the same or opposed directions, the entire velocity of the body is not added to or subtracted from the velocity of the light with respect to the body, rather it must be supplied with a reducing factor, the Fresnel entwinement coefficient. This served as a valuable touchstone for all later theories of the optics of moving bodies. The Fizeau experiment was long held to be a decisive proof of the existence of an ether, which was supposed to permeate all bodies without participating in their motion. Only in this way could this reducing factor be understood. It was reserved for the relativity theory to disprove this argument. It pointed out that the addition or subtraction of velocities, hitherto taken as self-evident, is not justified under the conditions prevailing here. Consequently, the history of the Fizeau experiment is an instructive illustration of the extent to which even theoretical elements enter into the explanation of every experiment; they cannot be excluded. Accordingly, if theories change, what has been an impressive proof of the truth of one of them can easily become an equally strong argument in favor of one that is quite different.

The old notion of the additivity of the velocities of light and moving bodies, however, found verification in other connections. For example, in 1842 Christian Doppler (1803-1853) deduced from the undulatory theory that the observed vibration frequency increases as the light source and the observer approach each other, and decreases as they get farther apart. Although it is difficult to understand why, it is a fact that this Doppler principle, despite its acoustical confirmation (Chapter II), was bitterly opposed for several decades. The reason, in part, stemmed from Dopplers untenable applications of it to astronomy. Yet he was correct in so far as astronomy was the first field in which substantiation of the principle was obtained. In 1860 Ernst Mach (1838-1916) stated clearly that those absorption lines in the stellar spectra that come from the stars themselves must show the Doppler effect; he likewise pointed out that these spectra contain other absorption lines of terrestrial origin which do not show this effect. The first pertinent observation seems to have been that by William Huggins (1824-1910) in 1868. Today the accuracy, under favorable conditions, of such observations is so high that radial velocities of 3×10^4 cm/sec can be detected, whereas some up to 10^7 cm/sec occur. The laboratory proof of the Doppler effect was obtained in 1905 by Johannes Stark. He worked with canal rays, i.e., luminous atoms, which in electrical gas discharge acquire velocities up to 10^8 cm/sec, so that the Doppler shifts of the spectral lines are incomparably greater than in astronomy. In 1919 Q. Majorans verified the effect by means of mechanically moved light sources with velocities of about 2×10^4 cm/sec.

No matter how great the importance of aberration and Doppler effect, they do not answer the question as to the existence of several optically justified reference systems. More intimate studies show that they are not concerned with the velocities of the light source and the observer with respect to a reference system, but only – at least in first approximation – with the relative velocity of the two with respect to each other. On the other hand, would the existence of a preferred reference system be proved, if an observation, with all the participating bodies moving at the same velocity, should reveal an effect on this velocity? Under such circumstances, this velocity would be in competition with the velocity of light; the ratio of the two velocities is involved, and this is always a small number. Consequently, such observations are difficult, even when they involve an effect of not more than the first order, i.e., proportional to this ratio itself, and are exceedingly so when an effect of the second order enters, i.e., when the square of the ratio is involved. It is hopeless to attain adequate velocities of all the participating bodies in the laboratory. The velocity with which the earth revolves around the sun is necessary for this purpose; but even in this case the ratio equals 10^{-4}. The object of such experiments is to prove this orbital velocity or, in other words, to demonstrate the ether wind with respect to the moving earth.

Attempts to do this have been made ever since 1839, when Jacques Babinet (1794-1872) sought to discover an influence of the earth's motion on interference phenomena. The results invariably were negative. Most of the experiments involved effects of the first order and could no longer be applied in deciding the question regarding the reference system after H. A. Lorentz in 1895 proved, by means of the electron theory, that there can be no such optical or even electromagnetic effects of the first order. Hence, the few experiments which involve effects of the second order became all the more significant. Among these none is easier to understand as to its underlying theory and more certain in its experimental accomplishment than the Michelson experiment. It gives a direct comparison of the relative velocities with which light travels in different directions with respect to the earth. The ether wind, if it exists, would have produced differences between these velocities.

The idea, and the first, though inadequate trial was published by A. A. Michelson in 1881. After Lorentz had pointed out its defects in 1884, Michelson and E. W. Morley repeated the experiment in 1887 with modifications that provided the required accuracy. In 1904 Morley and D. C. Miller advanced much farther; they reported that the observed effect did not amount to even one one-hundredth of the expected result. Although after 1920 Miller thought he had obtained positive results at great altitudes, they were contradictory among themselves and furthermore were disproved by repetitions carried out in 1926 by I. R. Kennedy and by several measurements made in 1926-27 by A. Picard and E. Stahel, some on Mt. Rigi and some in a balloon. The accuracy was enhanced so much by K. K. Illingworth in 1927 and by G. Joos in 1930 that an ether wind" of 1 or 1.5 km/sec must have made itself evident, if the theory of the preferred reference system were correct.

The influence of the Michelson and several other similar experiments gave rise to the special relativity theory. This ushered in a new era for the problem of the reference system. It asserts as a natural law the existence of an infinite number of inertial systems, which have translatory motion with respect to each other of constant velocities, and are equally justified for the *totality of natural events*. To be sure, when converting from one to the other, it is not possible to proceed as in Newtonian mechanics, namely, the time is not transformed at all and all material distances are left as they were; therefore the mechanics also requires the modification mentioned in Chapter II To express the fact that light is propagated with the same velocity in all directions in every inertial system, as indicated by the Michelson experiment, it is necessary rather to have a simultaneous transformation of the local coordinates and the time. This Lorentz transformation leads, for example, to the law that every body which is moving against an inertial system is shorter in this direction than when it is at rest. Admittedly, this contraction is slight, of the second order for low velocities; but when the velocity approaches that of light, the shortening becomes very important. Indeed, the measuring of the body in this direction must sink below every limit, when the speed of light is closely approached. Another consequence is that the velocity of light becomes the upper limit not only for all corporeal velocities and for the propagation of all physical effects through space, but also for the relative velocities of all inertial systems with respect to each other. Accordingly, the velocity of light goes beyond the bounds of optics and electrodynamics and attains universal significance for natural events. It was a sort of historic accident that humanity first discovered this in the case of light.

The fact that this transformation from one optically justified reference system leads to another which is equally justified was present, in essence, in a study made by Waldemar Voigt (1850-1919) in 1887. It was explained around 1900 through ingenious tentative ideas by Henri Poincaré. In 1904 it was confirmed, with the aid of electrodynamics, by H. A. Lorentz, who even then contributed the relativistically modified mechanics (Chapter II). Nevertheless, the viewpoint of all these forerunners remained that electromagnetic and optical processes occur as though this transformation leads again to a justified reference system. Lorentz, for example, differentiated in definite terms between the actual absolute time, which can be used directly for a justified reference system, and the local times, which can be calculated from it and the local coordinates for other reference systems. The decisive change, the omission of the as though, was made in 1905 by Albert Einstein. On the strength of a deeper insight into the essence of the measurement of space and time, he announced the complete equal validity of all reference systems derived by this transformation from a valid system, and therefore also the equivalence of all space and time measurements appertaining to them. The polemic against the relativity theory, in part, went beyond all reasonable bounds; it arose because many of the opponents lacked the requisite keenness of insight. This fundamental inversion then led its originator to the crowning glory of the whole structure of the relativity theory, namely, the law of the inertia of energy (Chapter II).

A material luminiferous ether is incompatible with the relativity theory; as was stated, it results in the preference for a particular reference system. With this, the Faraday-Maxwell concept of the electromagnetic field as a modified condition of the ether is eliminated; nothing remains except to regard this field itself as a reality.

Another conclusion drawn from the Lorentz transformation is that a clock in motion runs slower than when at rest. However, those periodic vibrations in atoms which produce the light of the spectral lines can be regarded as a clock. Of course, this effect is small, of the second order, and accordingly difficult to perceive. However, the velocities of the canal rays sufficed to reveal it. In observations of the Doppler effect, it superimposes itself on the familiar Doppler shift as a quadratic Doppler effect, and this was actually observed in 1938 by H. I. Ives and G. R. Stievel, and by H. Oiting in 1939.

The special relativity theory, which was discussed here, formed the close of a development which had been proceeding through a century. Precisely for this reason, it did not present experimental investigation with any new problems. Whatever new information has since been given by appropriate studies has come from improvements of earlier experiments. It has been pointed out to what extent these repeated the Michelson experiment, but it is useful to call attention also to an electrical-mechanical experiment devised and carried out in 1903 by T. Trouton and H. R. Noble, and refined in 1926 by R. Tomaschek. The total accuracy of the Michelson experiment was thus increased. According to the electronic theory, when a charge is given to a rotatable, suspended electrostatic condenser, it should turn because of the motion of the earth, whereas, according to the relativity theory, no rotation would ensue. The calculated effect would admittedly be slight, of the second order, but these workers were able to prove that it is not present.

The mathematicians and theoretical physicists found themselves with all the more to do. They had to adjust all branches of physics to the relativity theory, i.e., such fields as hydrodynamics, the elasticity theory, thermodynamics, and the parts of the Maxwellian theory that relate to matter. The relativity theory owes its elegant mathematical form to Hermann Minkowski (1864-1909) who, shortly before his death, introduced time as the fourth coordinate, with a validity equal to that of the other three. However, this addition involves nothing more than a very valuable artifice; it does not connote anything deeper,[3] even though some tried to read a more profound meaning into it. Max von Laue brought out the first comprehensive presentation of the special relativity theory in 1911.

Einstein did not stop with the special relativity theory. The conditions with respect to the measurement of space are like those surrounding the measurement of time (Chapter I). A continuum is presented to our intellect, and it is necessary to introduce a system of measures; but this problem is more involved because of the three dimensions of space. In principle there are an infinite number of equally valid methods for this. The mathematicians make the most of this multiplicity when they freely invent non-Euclidean geometries. Physics, however, is obliged to restrict this multifariousness because of the practical requirement that its geometry must make possible a simple presentation of the natural laws. This is the core of the question as to which geometry holds *empirically*. When, for instance, Gauss, in order to test the validity of Euclidean geometry, determined by geodetic, i.e., optical, methods, whether the sum of the angles of the triangle formed by three peaks (Brocken, Inselberg, Hoher Hagen) actually amounts to 180° as required by this geometry in contrast to the others, he tacitly required in interest of this simplicity that the rays of light follow geodetic ("shortest) lines. Whoever disclaims this can draw no geometrical conclusions at all from the experimental result confirming this sum of the angles.

Whereas physics, up to that time, had been able to get along perfectly well with Euclidean geometry, Einstein's general relativity theory, as it had gradually developed from 1913 on, believed that it would be forced to draw upon a non-Euclidean Riemann geometry. The deviations from the Euclidean are minimal, even in the vicinity of masses as large

---

[3]EDITOR'S NOTE: Unfortunately, this estimation of Minkowski's epoch-making discovery of the spacetime structure of the world is profoundly incorrect. See Editor's Note at the end of this chapter.

as the sun, and appear in only very few observations. These are (1) the advance in the perihelion of Mercury, which could not be explained by the planetary theory (Chapter III); (2) the change in direction of light close to the sun, which Arthur Stanley Eddington (1882-1944) found during the 1919 solar eclipse to correspond quite closely to that predicted by Einstein, while later eclipse observations, of course, produced a somewhat greater value. The third instance of the verification of the theory is drawn from the recent spectral studies of a particularly dense star, the companion of Sirius. The spectral lines from it are shifted considerably toward the red as compared with their position in a terrestrial spectrum. The case concerning this theory is not yet closed; but it will always bear the honor of having predicted the deflection of light without special *ad hoc* assumptions.[4]

EDITOR'S NOTE

I think, very reluctantly, that von Laue's inability or unwillingness to see "anything deeper" in Minkowski's foundational contributions to spacetime physics (p. 86) should be addressed.

If there were indeed nothing deep in Minkowski's discovery that the physical world is four-dimensional, i.e., if

---

[4]Others had predicted such deflections of light, but with such assumption.

spacetime (representing this world) were nothing more than a mathematical construction, then one should *first* ask why would a mathematician announce so excitedly[5] the introduction of one more mathematical space:[6]

> The views of space and time which I want to present to you arose from the domain of experimental physics, and therein lies their strength. Their tendency is radical. From now onwards space by itself and time by itself shall completely fade into mere shadows and only a specific union of the two will still stand independently on its own.

Then the decisive arguments come. The very essence of Minkowski's 1908 lecture is his successful decoding[7] of the profound physical message hidden in all failed experiments (the experiments captured in Galileo's principle of relativity and the Michelson-Morley experiment) to detect uniform

---

[5] Apparently Minkowski had realized the entire depth and grandness of the new view of the absolute four-dimensional world imposed on us by the experimental evidence. A draft of his Cologne lecture "Space and Time" reveals that he appears to have tried to tone down his excitement in the announcement of the unseen revolution in our understanding of the world. As the draft shows, Minkowskis initial intention had been to describe the impact of the new world view in more detail he had written that the essence of the new views of space and time is mightily revolutionary, to such an extent that when they are completely accepted, as I expect they will be, it will be disdained to still speak about the ways in which we have tried to understand space and time (quoted from: P. L. Galison, Minkowskis Space-Time: From Visual Thinking to the Absolute World, *Historical Studies in the Physical Sciences*, 10 (1979) pp. 85-121, p. 98).

In the final version of the lecture Minkowski had reduced this sentence about the new views of space and time to just Their tendency is radical.

[6] H. Minkowski, Space and Time, in: H. Minkowski, *Spacetime: Minkowski's Papers on Spacetime Physics*. Translated by Gregorie Dupuis-Mc Donald, Fritz Lewertoff and Vesselin Petkov. Edited by V. Petkov (Minkowski Institute Press, Montreal 2020), pp. 57-76.

[7] This naturally explains Minkowski's excitement when he introduced the new views of space and time (and therefore the new view of the physical world) in the beginning of his epoch-making lecture.

motion with respect to the absolute space (or the ether) –
Minkowski realized that all those experiments failed because
the measurements were, in fact, carried out in the space in
which the apparatus and the experimenters were at rest by
using the time associated with that space; that is why the
experiments always confirmed the state of rest of the apparatus and the experimenters in their own space. This was the
profound (totally counter-intuitive) physical message hidden
in the failed experiments to discover absolute motion.

Minkowski decoded that hidden message by analyzing[8]
the introduced by Lorentz local time t' (in addition to the
true time t of an observer at rest with respect to ether) in his
effort to explain the negative result of the Michelson-Morley
experiment (the failure to detect Earth's motion in the ether
or in the absolute space). Lorentz regarded t' as a purely
mathematical notion that did not represent anything in the
physical world. Einstein *postulated* that motion was relative,
which implies that time was also relative, and insisted that t
and t' should be treated on equal footing. Led by his analysis of the internal logic of the mathematical formalism of
classical mechanics, Minkowski also arrived at the conclusion
that t and t' should be treated equally. But, then, it was
exceedingly obvious to a mathematician that if two observers
in relative motion have different times they inescapably have
different spaces[9] as well (since space is orthogonal to the time
axis). And Minkowski explained how the existence of many
spaces reveals what he meant at the beginning of his lecture
(mentioned above):

> Hereafter we would then have in the world no
> more *the* space, but an infinite number of spaces
> analogously as there is an infinite number of planes
> in three-dimensional space. Three-dimensional

---

[8]Lorentz' local time played a key role in Minkowski's analysis of
the mathematical structure of classical mechanics; see the first part of
Minkowski's paper "Space and Time" mentioned above.

[9]After realizing this, Minkowski remarked: "Neither Einstein nor
Lorentz disputed the concept of space."

geometry becomes a chapter in four-dimensional physics. You see why I said at the beginning that space and time will recede completely to become mere shadows and only a world in itself will exist.

Indeed, the existence of many spaces implies a four-dimensional world because *many spaces are impossible in a three-dimensional world* (where there exists a *single* and therefore absolute space).

Minkowski did not state it explicitly (perhaps it looked too obvious to him), but neither the negative results of the experiments captured in Galileo's principle of relativity and the Michelson-Morley experiment nor length contraction would be possible if spacetime did not represent a real four-dimensional world.[10] Now we also now that, similarly, none of the kinematic relativistic effects are possible in a three-dimensional world.[11]

[10]V. Petkov, *Seven Fundamental Concepts in Spacetime Physics*, SpringerBriefs in Physics (Springer, Heidelberg 2021), Ch. 1.

[11]V. Petkov, *Relativity and the Nature of Spacetime*, 2nd ed. (Springer, Heidelberg 2009), Ch. 5.

7 THE BASES OF THE THEORY OF HEAT

Even pre-scientific experience had taught the difference between warm and cold objects and the equalization that ensues when objects with different degrees of heat are placed in contact with each other. It was known, in fact, that if an object A is in thermal equilibrium with two others, B and C, the latter are also in equilibrium with each other. This experience led, even prior to actual scientific study, to the arrangement of the degree of warmth according to a one-dimensional scale, i.e., to the creation of a qualitative concept of temperature, in which, of course, it was possible to speak only of high and low temperatures, without combining this relation with measure and number. The need of quantitative determinations of temperatures arose with the initiation of scientific investigation. Hence, Galileo Galilei, Evangelista Torricelli, Otto von Guericke, and many of their contemporaries tried to construct thermometers. The basis of all these was the thermal expansion of liquids or gases, just as in most of the present-day thermometers. Naturally, these early thermometers were subject to many disturbing influences, such as the atmospheric pressure, and thus yielded results of quite limited usefulness. Furthermore, because of technical difficulties, thermometers of the same construction did not show agreements in their readings. The first to overcome these defects and difficulties, and thus to become the father of thermometry, was Gabriel Daniel Fahrenheit (1686-1736), whose labors go back at least to 1709. His construction is the one most often used for the household thermometers of today. This was the first step toward a science of heat.

These thermometers accomplished the establishment of fixed points in the temperature scale, an advance somewhat similar to the discovery of the Fraunhofer lines in the spectrum (Chapter IV). To assign numbers to these points still remained a problem however; here there was a similarity to the measurement of time, despite the purely empirical character of the temperature concept. In both cases – just as in the spectrum – physics was confronted with a one-dimensional continuum and had to impress measure and number on it. Here again the guiding principle was furnished entirely by the goal of adapting this measuring system as well as possible to a simple formulation of the natural laws.

Without exception the early temperature scales were arbitrary. The use of thermal expansion as a means of measuring temperature was an arbitrary choice in itself; many other properties of materials were available for this purpose. High or very low temperatures in particular are now often determined by means of the electromotive force of thermocouples or the resistance of a bolometer wire. Even if thermal expansion was adhered to, the choice of the thermometric substance was arbitrary, no matter whether alcohol, mercury, or perhaps a gas was employed for this purpose. The arbitrariness is only partly removed if one of the ideal" gases is chosen. Determinations published in 1801 by John Dalton (1766-1844) and by Joseph Louis Gay-Lussac in 1802, which in 1842 Heinrich Gustav Magnus (1802-1870) and independently Henri Victor Regnault (1810-1878) confirmed with increased accuracy, showed that the thermal expansion of these gases is practically alike, and hence it represents at least more than the property of a single material. With respect to this problem, it is of subordinate importance whether the zero point of the scale is established by means of a definite freezing mixture, as was done by Fahrenheit, or whether it is placed at the freezing point of water, as was done by René Réaumur

(1683-1757) and Anders Celsius (1701-1744), and whether the other fixed point, the temperature of boiling water, is assigned the scale number 212, or 80, or 100. The solution of the problem was not obtained until 1854; it came out of the second law of thermodynamics, which will be discussed in Chapter IX. This law refers the natural temperature scale back to the measurement of quantities of heat.

The conceptual differentiation of quantities of heat from temperature is due to Joseph Black (1728-1799), who by this means soon after 1760 accomplished the second great advance in the theory of heat. In complete conformity with his thought, the unit quantity of heat, the calorie, is defined as that quantity of heat which raises the temperature of one gram of water one degree centigrade. Accordingly, this unit seems to be dependent on the measurement of temperature, but this is only apparently so. The discovery (1842) of the equivalence of heat and energy by J. R. von Mayer permits the determination of heat quantities in mechanical terms. Hence, the measurement of temperature is likewise fundamentally referred back to mechanical measurement. The classic instrument for determining quantities of heat is the ice calorimeter; it was described in 1780 by Antoine Laurent Lavoisier (1743-1794) and Pierre Simon Laplace. Black and J. C. Wilcke, independently of each other, had previously introduced the concepts of specific heat and latent heat which accompanies melting or vaporization.

The temperature definition of the second law of thermodynamics will be examined somewhat more closely. It deals with a reversible cyclic process, during which a body expands once isothermally with absorption of a given quantity of heat, then expands further without the gain or loss of more heat, then is isothermally compressed with liberation of heat, and finally undergoes, with no heat exchange with the surroundings, a further compression to the exact extent that it returns to its initial condition. The temperature definition states: The temperature of the two isothermal changes of state are in the same ratio as the quantity of heat added is to the quantity of heat given off. The fundamental law states that this ratio is independent of the nature of the body which is subjected to the cyclic process. In this way the temperature measure is defined except for a proportionality factor. The latter is so chosen that the temperature difference between the freezing and the boiling point of water amounts to 100°. This produces the absolute thermodynamic temperature scale; it was made the legal scale in Germany on September 8, 1924. Determinations showed that on this scale, the freezing point of water is at 273°. Within the range important to daily life, the readings on mercury or alcohol thermometers provided with the centigrade scale agree with it sufficiently well, apart from the difference of the zero points.

Experience has shown that the two quantities of heat, by means of which the temperature is defined, are always positive quantities. Consequently, there are no negative absolute temperatures; this scale has an absolute zero point. This could have been avoided by using a suitable function of this temperature, such as its logarithm, as temperature measure. This would be entirely possible; no natural law opposes it. When it is not done, it represents, as Ernst Mach correctly pointed out, the convention residuum in our temperature concept. If it should be done, the scale would extend as far as negative infinity, and it would avoid giving the impression on reaching 1° absolute, for instance – distinctly lower temperatures have been reached – that bodies cannot be cooled much farther. Actually, as Walter Nernst recognized in 1906, absolute zero is unattainable.

Since experience with temperature equalization taught that one body receives exactly as much heat as the other gives up, Black and his contemporaries believed heat to be a material substance, which could be neither created nor destroyed. Likewise, in connection with the steam engine, which was developed into a revolutionary economic factor by James Watt (1736-1819) around 1770, nobody realized at first that the heat delivered to the steam boiler is partly transformed into mechanical work, and therefore lost as heat. This error was the reason why, at first, no fruit was borne by Sadi Carnots (1796-1832) brilliant intuition that the output of steam engines is connected by a *universal* law with the passage of heat from a high to a lower temperature. It was only after the discovery of the equivalence of heat and energy, that Rudolf Emanuel Clausius (1822-1888) was able to derive from it the second fundamental law (Chapter IX). Even this fact reveals how great was the change in physics produced by the principle of the conservation of energy.

A quite different and by no means simple question is: How can the thermodynamic temperature scale be put into practice? The cyclic process used in its definition is a mental experiment which could hardly be carried out in any case with sufficient accuracy. However, the development of thermodynamics provides ways and means of converting other scales into the thermodynamic scale. Details cannot be given here, but it can be pointed out that for high temperatures the heat radiation is very successfully used for determining temperatures, especially since the temperature of the radiator is related to the radiation by simple and theoretically well-grounded laws. The temperature of the fixed stars is also measured in this way, an accomplishment which is of great importance, of course, to astronomy.

The earliest means of lowering temperatures were freezing mixtures and cooling by means of rapidly evaporating liquids. When their limits had been reached, these procedures were gradually replaced by the effect discovered in 1852 by James Prescott Joule (1818-1889) and William Thomson (later Lord Kelvin), (1824-1907). In this Joule-Thomson effect, a properly precooled gas is compressed and then allowed to escape through a narrow orifice. The slight cooling obtained by this expansion was developed in the nineteenth century, especially by Carl Linde (1842-1934), into a special cooling technique. Its characteristic feature is the counter flow or heat exchanger, in which the parts of the gas that have already expanded and so been cooled are used to precool the remaining gas which is still to be expanded. This process can be carried so far that the critical temperature is passed and part of the gas becomes liquid. In this way the permanent gases oxygen and nitrogen were liquefied in considerable amounts in 1883 by Zygmunt Florenty von Wroblewski (1845-1888) and Karol Stanislaw Olzewski (1846-1915). In 1898 James Dewar (1842-1923) succeeded in liquefying hydrogen, and in 1908 Heike Kammer-lingh-Onnes accomplished the momentous feat of obtaining liquid helium (Chapter V). The last permanent gas had been liquefied.

If one of these liquids is allowed to boil under reduced pressure, temperatures well below the liquefaction temperature are obtained. In this way, about $10°$ absolute was reached with hydrogen, and $0.7°$ absolute with helium.

The mathematical theory of heat conduction was founded on the concepts of temperature and indestructibility of the quantity of heat in 1804 by J. B. Biot and in final form by J. B. J. Fourier (1768-1830) in 1807 and 1811. The methods constructed for this purpose arc among the classic tools of mathematical physics. This is particularly true of the representation of arbitrary functions by series or integrals of the sine functions. Fourier resolution into pure sinusoidal vibrations plays an important part in the theory of every wave process, be it sound, surface waves on liquids, or electromagnetic oscillations, and all the more so as every acoustic resonator, every optical spectral apparatus, accomplishes this analysis automatically (to a certain degree). Subsequently, mathematics created the analysis into other systems of orthogonal functions, which are now of incomparable value in the solution of the Schrödinger equation (Chapter 14). Fourier's work is a model instance of the initiation of a fundamental advance in mathematics because of the needs of physics.

8 THE LAW OF CONSERVATION OF ENERGY

From the historical standpoint, the energy principle stems from mechanics. Even Galilei used it, less as a result of experiment than as an intuition, in the form that the velocity reached by a falling object was capable of raising it again to the original height but no higher. Huygens generalized this idea for the center of gravity of a system of falling bodies. In 1695 Leibniz gave it the form: the product of the force times the path equals the increase in "vital force' (*vis viva*). Newton attached no particular importance to this notion. On the other hand, Johann Bernoulli (1667-1748) speaks repeatedly of the *conservatio viritim vivarum* and emphasizes that when vital force disappears the ability to do work has not been lost, but is merely changed into other forms. Leonhard Euler (1707-1783) knew that if a mass point moves under the influence of a central force, the vital force is the same each time the point reaches a certain distance from the center of attraction. By 1800 ripened experience had established the principle that in a system of mass points which exert central forces on each other the vital force depends only on the configuration and its dependent force function. The designation energy for vital force was coined in 1807 by Thomas Young, the term work by Jean Victor Poncelet (1788-1867) in 1826.

Accordingly, the impossibility of constructing a perpetuum mobile by purely mechanical means was established. By the end of the eighteenth century it was also the general conviction that it could not be accomplished by any other means either;[1] at least in 1775 the French Academic resolved no longer to consider alleged solutions of this problem. However, the positive gain to science that came from this very negative sounding insight did not become apparent until the nineteenth century.

The first scientist to connect heat and work was Sadi Carnot, whose effort was thwarted by the erroneous belief that heat, with respect to quantity, consists of an unchangeable substance (Chapter VII). Unfortunately an essay left by the prematurely deceased Carnot was not published until 1878, when the energy principle had long since been accepted. In this posthumous paper he abandoned his earlier view and, without derivation, states a mechanical equivalent of heat which was fairly correct. Of course, it came too late to have any influence on the course of history.[2]

Even the ancients knew that the temperature of objects rises when they are rubbed; the material theory of heat sought to explain this fact by all sorts of hypotheses about friction. Benjamin Thompson, later Count Rumford (1753-1814), demolished these notions in 1798 when he pressed the dull end of a borer against the bottom of a cannon barrel and set it turning by horsepower. In this way, even considerable quantities of water were brought to boiling, without the heat capacity of the metal showing any of the change demanded by these hypotheses. The same point was made in 1799 by Humphry Davy when, by means of a clock work, he caused

---

[1] EDITOR'S NOTE: This can be regarded as a kind of precursor of a similar discovery of a fundamental feature of the world – what we call Galileo's principle of relativity captured the impossibility of detecting uniform motion with respect to the absolute space by purely mechanical means; in 1887 Michelson and Morley demonstrated that absolute motion cannot be detected by electromagnetic means, but hardly in his 1905 paper Einstein *postulated* that it could not be accomplished by any other means either and in 1908 Minkowski *explained* why (see Editor's note at the end of Chapter 6).

[2] See in this connection, M. Planck, *Das Prinzip der Erhaltung der Energie*, Leipzig and Berlin, 1908, 2nd ed., p. 17.

two pieces of metal to rub against each other in a vacuum. The suspicion of the existence of a force which, according to the circumstances, appeared as either motion, chemical affinity, electricity, light, heat, or magnetism, so that each of these phenomena is convertible into the others, was voiced frequently in the first decades following 1800. To convert this suspicion into a reality, it was necessary to find a common measure of this force. Steps in this direction were made by various investigators, each in his own way.

The earliest of these was (Julius) Robert von Mayer (1814-1878), a physician, who in accord with his whole intellectual tendency, preferred to generalize philosophically rather than to build up empirically step by step.[3] Hence, in his short essay of May, 1842, Mayer applied the propositions "Ex nihilo nihil fit" and "Nil fit ad nihilum" to the "falling force," motion, and heat. The permanent residuum of this discussion was the fairly correct statement of the mechanical equivalent of heat. Of course, Mayer did not state how he arrived at this value until 1845; his calculation is the familiar one based on the difference between the two specific heats of ideal gases. It involved the assumption, not stated in his work, but actually provided by the measurements made in 1807 by Gay-Lussac, that the energy of such gases is not dependent on the volume. Ludwig August Golding (1815-1888) arrived at practically the same value by his own frictional experiments; his basis for the general conservation law seems now to be even more fantastic than that of Mayer. Even the latters second publication takes up electrical and biological processes, and in the third (1848) he explains the incandescence of meteors as due to their loss of kinetic energy in the atmosphere, and also applies the conservation law to the ebbing and flowing tide. Nevertheless, he was ignored at first, and the recognition which he so richly deserved came to him much later.

No matter what position is taken with respect to Mayers deduction, the following must be acknowledged in any

---

[3]M. Planck, *ibid.*, pp. 23, 24.

case: since the objective of physics is to discover general natural laws and since one of the simplest forms of such a body of laws is obtained when it expresses the immutability of a certain physical magnitude, the search for constant quantities is not only a legitimate but also a highly important field of investigation. It has always been represented in physics. Fundamentally, it is responsible for the early conviction of the constancy of electrical quantity. Of course, only experience can decide the question as to whether a magnitude regarded as constant is really unchangeable. The energy principle, like the law of conservation of electricity (Chapter V), is also an experimental principle. However, Mayer actually followed the empirical path when he calculated the mechanical equivalent of heat. As to other provinces of physics, the law for him was primarily a program whose accomplishment was left to others.

The second person of importance in this connection was James Prescott Joule, who in 1843 came forward with a paper (not published until 1846) dealing with the thermal and chemical actions of the electric current. By actual measurements, he proved that the heat developed in the connecting wire of a galvanic battery is equal to the heat effect (as it is now called) of the chemical reaction in the cell, provided the reaction proceeds without production of current,[4] and that this heat decreases when the current does work. Shortly thereafter and in 1845 he published determinations of the mechanical equivalent of heat in which he converted mechanical work into heat, partly directly, partly through electrical means, and partly through compression of gases.

However, the man whose universal mind was able to develop fully the universal significance of the conservation principle, was Hermann von Helmholtz (1821-1894). Like Mayer, of whose work he at first was not aware, and whose results he therefore achieved independently, Helmholtz came to the

---

[4]Obviously, this is true only if the battery undergoes no temperature change during the production of the current and exchanges no heat with the surroundings.

principle by way of medicine. In 1845, in a short paper, he corrected a slight error of the famous chemist Justus von Liebig (1803-1873) by pointing out that the heat of combustion of foods in the animal body may not be taken directly as equal to the heats of combustion of the chemical elements of which these foods are composed. At the same time, he gave a short synopsis of the consequences of the principle as it applied to various parts of physics.

This train of thought was then developed at greater length in his lecture before the Physical Society at Berlin on September 23, 1847. Helmholtz (in contrast to Mayer) based his arguments on the possibility of providing a mechanical explanation of all natural processes through central forces of attraction or repulsion, a method used by most of his contemporaries. He saw in this a sufficient and – erroneously – also a necessary condition for the impossibility of perpetual motion. However, he made no use of this assumption in his deductions, but rather derived the various expressions for energies directly from this impossibility, if for no other reason than the total failure of attempts to refer all processes back to mechanical forces. His propositions were therefore not attached to this idea and were able to outlive it. New, in contrast to his predecessors, were his concepts: potential energy" for mechanics, also the energy expressions for gravitation, for static, electrical, and magnetic fields, and also new was what he said concerning the energetics of the production of current by galvanic and thermoelements, and also regarding electrodynamics including induction phenomena. The modern method of calculating the energy of a gravitational field from the products of the masses times the potentials, and that of an electrostatic field by multiplying together charges and potentials, is based directly on Helmholtz.

It would lead too far afield to go into more detail; likewise the further development of the principle cannot be discussed here. Only the final definition which was given in 1853 by William Thomson (Lord Kelvin) will be mentioned. He stated that the energy of a material system in a given state can be designated by the sum of all effects, measured in mechanical equivalents, produced outside the system, when it passes by any way whatsoever from the given state to an arbitrarily fixed null state. The natural law of the conservation of energy resides in the words by any way whatsoever."

The Helmholtz statements of 1847 by no means found general agreement at once; his older contemporaries feared that they contained a revival of the fantastic notions of the Hegelian natural philosophy, against which they already had had to battle so long. Only the mathematician Gustav Jacob Jacobi (1804-1851), who rendered such excellent service to mechanics, immediately saw that they were the legitimate continuation of the course of thought of those mathematicians of the eighteenth century who had so greatly developed mechanics. When, then, around 1860 the energy law received general acceptance, it of course very quickly became a cornerstone of all natural science. From then on, and in physics especially, every new theory was examined first of all to see if it accorded well with the conservation principle. In fact, about 1890 some became so enthusiastic about this principle that they even went so far as to make it absolutely the central point of a *Weltanschauung*, namely, "energetics," or at least they attempted to derive all other physical laws from it. For instance, they so greatly misconstrued the second law of thermodynamics that they denied the difference between reversible and irreversible processes and, for example, placed the transfer of heat from high to lower temperatures on the same plane as the falling of bodies in the gravitational field. Max Planck was not very successful in combating this movement from the standpoint of thermodynamics, but Ludwig Boltzmann made more progress on the basis of the atomic theory and statistics. Like many other errors, energetics finally passed from the scene because of the death of its advocates.

The energy concept also entered into technology, where every machine is appraised on the basis of its energy balance, i.e., the ratio of the energy furnished to the output of energy in the desired form. The concept is part of the mental equipment of every scientist today.

The theory of energy was by no means completed by the acceptance of the law of conservation; rather it is still maturing by virtue of new developments. As stated, Helmholtz calculated the energy of electrostatic or electromagnetic fields from the charges and potentials. The application of Faradays idea of action at close hand led Maxwell to localize this energy in space, and to assign its own share to each element of volume. These ideas were carried further in 1884 by J. H. Poynting (1852-1914) who, for changing fields, in which the volume elements did not retain their quotas, developed a theory of energy flow just as though electromagnetic energy were a material substance. In 1898 G. Mie showed how this conception can be applied also to elastic energy. For instance, a current of energy flows in the opposite direction through the driving belts, which connect a steam engine with a machine, and if the connection consists of a rotating shaft, the energy flows in it parallel to its axis. Closely related to this idea is Plancks extension (1908) of the Einstein principle of the inertia of energy (Chapters II and VI). It states: an impulse (in the mechanical sense) is associated with every energy flow. The density of the impulse, i.e., the impulse per unit volume, is obtained by dividing the energy-current density by the square of the velocity of light. In fact, it is known from the experiment, first carried out in 1901 by Peter N. Lebedew (1866-1911), concerning the pressure exerted on bodies by light or other electromagnetic radiation that this pressure supplies the impulse. Around 1900 Henri Poincaré, H. A. Lorentz, and others had also set up this principle with restriction to electromagnetic energy.

According to Newtonian mechanics, there is a special kinetic energy; it is added to all other types of energy as a consequence of motion. The relativity theory eliminates this type of energy; instead, motion increases every type of energy by a factor that depends on the velocity. This fundamental change in viewpoint is closely connected with the principle of inertia of energy. A vicious circle would result, if on one hand, an energy form should be attributed to the inertia of bodies, and on the other, if every inertia should be ascribed to energy.

The principle of inertia of energy is used less often in the version just given than in the following form: the mass of a body is equal to its energy (at rest) divided by the square of the velocity of light. This imposes a restriction on the law of the conservation of mass. Addition of heat or work, e.g., compression of the body, increases its mass; loss of heat or work decreases it. Chemical reactions, in so far as they proceed with the production of heat, lessen the total mass of the reactants, but under all circumstances the loss is so slight that the decrease is not revealed by even the best of balances. Consequently, Landolt (Chapter II) could find no trace of this change. However, in the transmutation of atomic nuclei, energy is released in amounts which are incomparably greater in proportion to the mass. They play a very fundamental part in nuclear transmutations (Chapter XI).

The principle of inertia of energy thus closes a gap that was still present in the foregoing definition, which employed an arbitrarily chosen condition as the null point of energy. If, however, every inert mass is referred back to energy, then the latter is immediately defined with the former without such arbitrariness. That such reference corresponds to nature is impressively demonstrated in the case of the electron and positron, which have been found capable of changing themselves completely into radiant energy (Chapter XIV).

Disregarding the tides and their energy, until recently all the energy known on the earth was really derived, in the last analysis, from solar radiation. Hence the acceptance of the law of the conservation of energy made an important issue of the question as to the source of the energy which is continuously radiated by the sun and stars. R. v. Mayer's notion that it comes from the kinetic energy of the meteorites that are continuously bombarding these great heavenly bodies proved inadequate when subjected to examination. In

1854 Helmholtz directed attention to the gravitational energy of the great nebular sphere, whose condensation, according to the Kant-Laplace cosmogony, had given rise to the sun and planets; a process, which actually would have converted this energy into other forms. But even such a store of energy would not have sufficed to supply the radiation of the stars through the billions of years of their proved existence. It remained for atomic physics (Chapter XI) to uncover a sufficiently ample source of energy. The high temperatures in the interior of the stars make possible reactions between atomic nuclei, reactions which can only be brought about in the laboratory by means of accelerated corpuscles. In 1938-39, C. Fr. v. Weizsäcker and H. A. Bethe made it possible that attention should be primarily directed, in this connection, to the union of protons and electrons; of which 4 and 2, respectively, combine to produce one helium nucleus. This process does not proceed directly, but it can occur via several well-known nuclear reactions.

Nuclear transformations can provide mankind with energy directly, though at present in not more than extremely modest quantities. Nevertheless, this advance provides an avenue of escape from the complete dependence on solar radiation, which up to now, either directly or indirectly, has been the sole available source of energy.

9 THERMODYNAMICS

Classical thermodynamics, formerly called mechanical heat theory, is based on three fundamental laws. The first is the law of the conservation of energy (Chapter VIII), especially the declaration contained in it that heat is a form of energy and consequently its quantity is measurable in mechanical units. The whole content of this law is embodied in the principle of the impossibility of perpetual motion of the first kind.

The second fundamental law declares the impossibility of perpetual motion of the second kind, i.e., the construction of a periodic machine whose sole effect is to convert heat into mechanical work. If it existed, it would be possible to bring heat continuously and without any other change in the participating bodies from a lower to a higher temperature by converting the heat into work at the lower temperature, and then reconverting the work into heat at a higher temperature, a process that would go on directly in such a machine. Even Sadi Carnot had appreciated that an uncompensated transformation from a lower to a higher temperature cannot be realized in any fashion, even indirectly (Chapter VII). His mistaken belief that heat is an unchangeable material substance was corrected by the first law. The latter cleared the path along which Rudolf Emanuel Clausius in 1850 and William Thomson (Lord Kelvin) in 1854 advanced to the second law. Just as the first law introduced a function of state, namely, energy, the second, in the form given to it in 1865 by Clausius, likewise gives rise to a new concept. He called this function entropy. However, while the energy of a completely isolated system remains constant, its entropy, made up of the sum of the entropies of its parts, increases with every change. The limiting case, in which the entropy remains unchanged, though always important as regards theory, is only an ideal, since it is never strictly attained in practice. Decrease of entropy likewise is forbidden by the laws of nature, even as a mental experiment.

All processes are thus separated into two classes. If a particular process that is accompanied by an increase in entropy could be reversed, either directly or indirectly, this would entail a decrease in entropy. Consequently, the process is actually irreversible. Of course, it is possible to conceive of reversible processes, i.e., those in which entropy is conserved. In a reversible cycle of operations, i.e., one consisting of perfectly reversible single processes, such as introduced into physical theory by Benoit Paul Emile Clapeyron (1799-1864) in 1834, and represented approximately by the steam engine, there are two isothermal and two adiabatic branches, the latter occurring without the addition or delivery of heat. The relation of the quantities of heat supplied or taken away on the isothermal branches depends only on the temperatures at which these branches lie. The definition of temperature discussed in Chapter VII uses this fact. The difference between the two quantities of heat gives the (positive or negative) work output, which therefore depends – if the cyclic process is carried out reversibly – likewise in its relation to one of the quantities of heat only on the two temperatures. Hence, the efficiency of a machine of this kind is only a question of the available temperature difference. Other things being equal, the efficiency is less in irreversible cyclic processes. These are some of the prominent ideas taken from the beginnings of thermodynamics.

The mere fact of the general existence of two mutually independent functions of state such as energy and entropy, enables mathematical analysis to make an abundance of assertions concerning the thermal behavior of bodies. The conclusion that every equilibrium in a closed system must correspond to a maximum of entropy proved to be still more important. As soon as the entropy function of various substances can be stated, it is possible to make declarations about the equilibrium between them. Thus even Clausius was able to state the theory of the equilibrium between different states of aggregation of the same material. The thermochemical theory of equilibria brought perspective and order into the boundless multiplicity of chemical reactions after August Horstmann (1842-1929) had applied the two fundamental laws to a special case of this kind in 1873. Jacobus Henricus vant Hoff (1852-1911), Josiah Willard Gibbs, and Max Planck especially distinguished themselves along this line. In 1882 H. v. Helmholtz took part in the development. The ancient idea of chemical affinity, which never was rightly understood, was successfully referred back to differences in energy and entropy; it was shown that affinity depends not only on the nature of the reacting materials, but also on the temperature and pressure. Thermodynamics was carried over into the theory of elasticity, and magnetism, since events in all these fields are usually also connected with heat phenomena. In short, there is really no province of physics in which thermodynamics has nothing to say. The mere fact that it is left out of consideration is in itself a sign of idealizing a situation.

The definitions of energy and entropy were incomplete at first, in so far as both functions of state could be calculated only with respect to an arbitrarily chosen initial state. The law of the inertia of energy fills this gap for them. The third fundamental law, which Walter Nernst formulated by gifted intuition in 1906, acts as a supplement here for entropy. In the form which was soon given to it by Planck, it states that the entropy of a chemically uniform material approaches zero when its temperature is brought close to the absolute zero. Nernst connected the third law with certain observations concerning the heat effects of chemical processes, and his proof, which at first was met with some justified criticism, has since been confirmed by accumulated experience. Among the conclusions to which it leads, there is, for example, the disappearance of specific heat and coefficient of expansion as the temperature null point is approached. Most important of all is the possibility based on it of theoretically predicting chemical equilibria in complete detail from purely thermal measurements, namely, specific heats. The numerous determinations of specific heats, which Nernst made especially at the lowest attainable temperatures by methods he devised particularly for this purpose and which were applied either by himself or by others under his direction, were used primarily for this purpose. It should be noted that the fertility of the Nernst heat theorem has by no means been exhausted.

The province of classical thermodynamics has been outlined above. Its limits are set by the fundamentally irreversible processes which are far from equilibrium, because the second law provides no equation but only an inequality for them. When calculating entropy, modern physics makes frequent use of statistical methods, which will be discussed in Chapter X. The true significance of the entropy concept, which M. Planck especially championed from the start of his career, appears precisely here. If desired, this concept may be avoided in classical thermodynamics; a suitable cyclic process can be invented for each special case, and thus the general consideration leading to this concept can be repeated in the specific instance. On the other hand, it is indispensable in thermodynamic-statistical methods. It played an important or even decisive part in the discovery of the Planck radiation law.

10 ATOMISTICS

The concept and name "atom originated in antiquity. It is, of course, difficult for the scientist to understand the role that the concept actually played in the thinking of Democritus (460P-371? B.C.) and his followers; in any case, they did not combine it with any observations. The voluminous literature of atomistics extends through all centuries, but so far as it antedates 1800, and despite the famous names that occur in it, no favorable verdict can be rendered concerning its usefulness. The only exception is the quickly forgotten statement[1] (1738) made by Daniel Bernoulli (1700-1782) concerning a kinetic theory of gases. Similarly, according to the testimony of Helmholtz the publications along this line in the first half of the nineteenth century justified, to a certain extent, the aversion to all theories shown, for instance, by the eminent experimenter H. G. Magnus and many of his contemporaries. The modern concept of atom and molecule was created by chemistry; how this came about belongs to the history of that science. Three achievements which physics could simply take over around 1850 were the following: (a) John Dalton (1766-1844) is chiefly responsible for the idea that the atoms of a chemical element are completely alike in all their properties; (b) the definition that the mass of an atom relative to the mass of the hydrogen atom is expressed by the atomic weight; (c) in 1811, Amadeo Avogadro (1776-1856) added the rule, which bears his name, that ideal gases at the same temperature and pressure contain equal numbers of molecules per

---

[1]In his great paper *Hydrodynamics*.

unit volume.

Apart from the idea of 1824 of L. A. Seeber (see Chapter XII) concerning crystal structures, the earliest form of physical atomistics was the kinetic theory of gases. Around 1850, the leading authorities at least acknowledged the equivalence of heat and energy; it suggested that heat could be regarded as molecular motion. On the other hand, the experiments by J. L. Gay-Lussac in 1807, and similar determinations in 1845 by J. P. Joule, demonstrated that the internal energy of ideal gases is independent of their volumes, which again showed the extreme weakness of the forces acting between their molecules. Thus in 1856, August Karl Kronig (1822-1879) and in 1857 Rudolf Julius Emanuel Clausius were somewhat forced to the idea of ascribing rectilinear motion to gaseous molecules, except during the instants when they are colliding with each other or striking the wall of the confining vessel. The law of impulse then immediately showed that the pressure of the gas is proportional, with a universal factor, to the mean kinetic energy of the molecules. It clearly followed from the Boyle-Mariotte-Gay-Lussac law that this energy is proportional to the absolute temperature, a very fundamental fact, which is not limited to gases, and according to modern quantum theory is subject to significant exceptions only at very low temperatures. A reliable computation of molecular velocity resulted at the same time. The value, 1.9×10^5 cm/sec, obtained for hydrogen at 300° absolute was of course unexpectedly high, and at first seemed to be incompatible with the slowness of diffusion of gases into each other, or with their low heat conductivity. The first direct determination, by O. Stern, did not come until 1920. However, in 1858 Clausius showed that these processes depend less on the velocity than on the mean free path between two collisions. Actual values of this path length were given in 1860 by J. Clerk Maxwell, on the basis of his own determinations of internal friction. In this same paper he freed the calculations of the gas theory from the temporarily adopted assumption that all molecules have the same velocity, and announced the law of the distribution of velocities. He himself furnished the proof of this law, which bears his name, but it was improved primarily by Ludwig Boltzmann in 1868. Although at first it was not accessible to experimental verification – the opposing difficulties were not overcome until 1932 by O. Stern – it nevertheless soon became the starting point of numerous generalizations whose consequences could be verified by measurements (see below). This, of course, redounded to its own advantage.

Information concerning the size and number of gas molecules was likewise obtained at this time. By assuming that the molecules of the simplest gases are spherical, Joseph Loschmidt (1821-1895) estimated their diameters from the mean free path and the volume occupied by one mol in the liquid state. He found the proper order of magnitude for the radii (10^{-8} cm) and for the number of molecules in one mol (10^{23}). This number, whose value is now known much more accurately, is called the Loschmidt or Avogadro number.

At that time, all studies of the gas theory, among others the proofs of the Maxwellian distribution of velocities, were based on this same assumption of spherical rigid molecules. Gradually, however, the theory was extended to molecules with internal degrees of freedom, rotations and oscillations of the atoms with respect to each other. A generalized distribution law was set up to deal with these, and from it was derived the most important consequence of the law of uniform distribution: The mean kinetic energy of every degree of freedom is proportional, with the above-mentioned universal factor, to the absolute temperature. This led to the direct calculation of the specific heats of polyatomic gases, and the results were in closest agreement with the experimental values. In fact, the application to solid bodies explained forthwith the law discovered in 1820 by Pierre Louis Dulong (1785-1838) and Alexis Therese Petit (1791-1820), namely, that the specific heat referred to the gram-atom (mol-heat) has a value, 6 cal/degree, that is common to all. Hand in hand with this went the answer to the question as to how the gaseous molecules distribute themselves in space under the influence of external forces, such as gravity. All these were basic facts, which later came to light in various fields.

Thus the foundations of the kinetic theory of gases were laid down. No changes in them came from the studies of M. Knudsen, who, taking advantage of the progress in vacuum technique, studied the special phenomena that arose in systems so highly evacuated that there were scarcely any collisions between the gas molecules. The principles are still valid, and the important theoretical investigations of thermodiffusion by D. Enskog (1911) and S. Chapman (1917), as well as the experimental discovery, in this same year, of this effect by S. Chapman and P. W. Dootson, together with the discovery by K. Clausius and L. Waldmann of the concomitant reverse effect, i.e., the heat phenomena associated with the diffusion of two gases, agree fully with the foundations that had been laid by Clausius, Maxwell, and Boltzmann.[2]

These foundations go back to Newtonian mechanics, and yet a new viewpoint entered physics with the kinetic theory of gases, namely, probability considerations. Any attempt to determine the zigzag path of every individual molecule from collision to collision would not only be a hopeless undertaking, but it would also have no scientific value. However, the *mean* free path, the *average* number of collisions which a molecule experiences in unit time, etc., are of importance. Pressure and temperature are *average* values, to be established for many molecules. No one appreciated the significance of this aspect of the theory more plainly than Planck, who condensed it into the hypothesis of molecular disorder. In this there resides an advantage of the Boltzmann method over the statistical mechanics of J. W. Gibbs, which is applicable not simply to gases, and which is sometimes easier to handle and likewise leads to equipartition laws. By this means Boltzmann was able to incorporate into the gas theory the fundamental difference between purely mechanical and

---

[2]Thermodiffusion in liquids had been observed as early as 1856 by Carl Ludwig (1816-1895) and in 1889 by Charles Soret (1854-1904).

thermal processes, which was often raised as an objection to every kinetic theory. Mechanical processes are basically reversible; each can proceed exactly as well in the opposite direction; the sign of the time factor plays no role. Thermal processes are fundamentally just as irreversible (Chapter IX) as the equalization of two different temperatures. If, despite its foundation in mechanics, the gas theory presents these and other processes also as irreversible, the reason lies precisely in the interjection of the hypothesis of molecular disorder. The analogy to the principle of the increase of entropy is obvious. Therefore, the relation between entropy and probability, one of the profoundest thoughts of physics, which Boltzmann clarified more and more from 1887 on, forms the capstone of his life work. This Boltzmann principle states: Entropy is proportional to the logarithm of the probability of state. The proportionality factor is known as Boltzmanns constant. Its value was not stated until 1900 by Planck (Chapter XIII). The increase in entropy, expressed by the second law of thermodynamics, thus becomes transformation into an ever more probable state. Since states of not much less probability are always near the state of maximal probability, temporary though slight fluctuations from it will occur. This is the important new fact. These thermodynamic fluctuations account for the permanent motion of ultramicroscopic particles suspended in liquids or gases. This verified molecular movement, discovered in 1827 by the botanist Robert Brown (1773-1858), and hence called Brownian motion, was long thought to be a thermal phenomenon. Its statistical theory, which was given in 1904 by Maryan von Smoluchowski (1872-1919) received its probably final form from Albert Einstein. These and many other oscillation phenomena constitute one of the most convincing proofs of atomistics, and they have brought about the conversion of numerous skeptics.

Independent of the gas theory, atomistics overlapped into the theory of electricity. After 1834, when Michael Faraday discovered the law of electrochemical equivalence, which states that a gram molecule of univalent ions, no matter what their nature, transports a definite electric charge with it, and that this charge is twice as large for divalent ions, etc., more than one physicist had the idea of assigning to each ion an electrical elementary charge, or its double, and so forth. This was done, for example, by Svante Arrhenius in his theory of electrolytic dissociation (1882), to which Walter Nernst added his brilliant theory of diffusion in electrolytic solutions and electromotive force of galvanic cells. Similarly, J. Larmor and H. A. Lorentz, in the electron theory (Chapter IV), assigned one elementary quantum apiece to the carrier of the electrical charge in matter. It was not until 1890 that the name electron to designate the negative unit charge was introduced by Johnstone Stoney (1826-1911).

This atomistics became especially important with respect to the manifold phenomena attending the discharge of electricity through gases. In 1859, Julius Plucker (1801-1868) discovered the rays, now called cathode rays, a name suggested in 1876 by Eugen Goldstein (1850-1930). Johann Wilhelm Hittorf found in 1869 that they can be deflected by a magnetic field, and finally, in 1871 Cromwell Fleetwood Varley (1828-1883) discovered their negative electrical charge. Then, in 1876 Goldstein spoke of their deflection in an electrical field; however, neither he nor Varley convinced physicists in general, so that it was not until 1895 that Jean Perrin (1870-1942) and in 1897 that Joseph John Thomson (1856-1910) decided the question in favor of Varley and Goldstein. Nonetheless, supported by the sensation created by the brilliant (1879) experiments of William Crookes (1832-1919), the idea gained ground that the cathode rays are corpuscular in nature, even though Heinrich Hertz, on the basis of a series of experiments (1883) which misfired because of inadequate experimental technique, believed them to be a longitudinal wave radiation. Canal rays, the counterpart of cathode rays, were described by Goldstein in 1886. On the basis of deflection measurements, Wilhelm Wien in 1898 drew a conclusion concerning the ratio of mass to charge in the case of these canal rays, just as its magnitude is calculated for electrolytic

116

ions from the Faraday equivalence law. On the other hand, in 1897 several investigators, including besides W. Wien and J. J. Thomson also George Fitzgerald (1851-1901) and Emil Wiechert (1861-1928), found the ratio of the mass of the cathode ray particles to their charge to be about 2000 times as small as that of the hydrogen atom. With final abandonment of the Hertz idea, it was concluded that: the particles of the canal rays are normal, electrically charged atoms or molecules, whereas the particles of the cathode rays are atoms or elementary units of negative electricity, i.e., electrons. Since, at the close of 1896, the Lorentz theory of the Zeeman effect (Chapter IV) for the charge-bearers emitting the spectral lines of the atoms had led to the same ratio of charge to mass, the existence of electrons was definitely established after forty years of effort. This result was confirmed in 1899 by E. Wiechert, who, by means of electrical oscillations, made direct measurements of the velocity of cathode rays and obtained agreement with the values derived from deflection experiments.

The most striking feature of the electron was its ability to penetrate considerable thicknesses of solid material. This had been observed as early as 1892 by Hertz, and in 1893 Lenard allowed electrons to escape from the discharge tube into the air through the "Tenard window." The subsequent research dealt especially with the absorption and dispersion of the electrons in matter; it is not yet completed. An especially powerful aid was provided in the cloud chamber devised by C. T. R. Wilson in 1912, which makes the paths of charged particles in gases, and therefore also of electrons, directly visible. As early as about 1900, Lenard developed his dynamid" theory of bodies to account for penetrability; it has many features in common with the later Rutherford model of the atom. About the same time, and especially under the direction of J. J. Thomson, a full explanation of the conductivity of electricity through gases also was developed; ions of both sign, and free ions besides, are its causes.

Many investigators from 1897 on tried to determine the

charge of the electron in absolute terms, and not merely in relation to its mass. The order of magnitude was known from the beginning, since it had been obtained from the accurately known charge of one mol of univalent electrolytic ions and the known order of magnitude of the Loschmidt number. However, the numerical values obtained at first were usually only two-thirds of the correct figure. The review of the gradual rise in the result creates a strange impression. Among the direct methods of measuring this charge, the best is now considered to be the procedure designed in 1907 by F. Ehrenhaft, used in 1913 by R. A. Millikan, and improved in 1940. The method employs an oil droplet carrying a few elementary charges and suspended in an electrical field. Its result, 4.796×10^{-10} electrostatic units, incidentally definitely confirms the belief that no smaller charges, i.e., no "subelectrons' exist, though there was widespread doubt that this was so. However, there are many indirect determinations of this value, since the elementary quantum is closely associated with the Loschmidt number on one hand and with the Boltzmann constant on the other. Measurements by means of X-ray interferences (Chapter XII) gave the value 6.0227×10^{23} for this number and accordingly 4.803×10^{-10} electrostatic unit for the elementary charge. The difference between the determinations of the charge is less than .2 per cent. From the historical standpoint, it is interesting to note that Planck in 1900 computed the value 4.69×10^{-10} from his radiation law and the then current radiation measurements. This value was much higher than the other contemporary determinations and it is now known that it was far superior to them in accuracy. More recent radiation measurements would raise his value to 4.76×10^{-10} but its accuracy is not comparable with that of the two other figures.

The indivisibility, from which the atom derives its name, holds for chemical changes and also for those collisions with other atoms that are treated in the kinetic theory of gases. However, investigators with deeper vision had often pondered the possibility that the atom might be made up of still smaller particles. In 1815, William Prout (1785-1850) reasoning from the fact that atomic weights were whole numbers, concluded that hydrogen was the common fundamental material of all atoms. However, improved methods of determining atomic weights during the nineteenth century left no doubt as to such wide deviation from this whole number relationship that Prouts hypothesis gradually sank into oblivion. Nevertheless, the idea of an inner connection between all elements was revived when Dmitri Ivanovitch Mendeleev (1834-1907) and Lothar Meyer (1830-1895), independently in 1869, arranged the elements into a periodic system purely on the basis of chemical behavior, a most brilliant feat whose full magnitude was not revealed until forty years later. In 1911 Ernest (later Lord) Rutherford, devised his atomic model in order to explain the scattering of α-rays in matter. In this, a planetary system of electrons surrounds a positively charged nucleus, which, though small, nevertheless carries practically all the mass of the atom. In 1913 Hans Geiger (1882-1945) and E. Marsden studied the deflection of α-particles over large angles, and later the X-ray spectra were investigated by Henry George Jeffreys (1887-1916) and others. These studies proved that the place of an element in the system, i.e., its atomic number, states the number of elementary charges which the nucleus bears. The periodic system is simply an arrangement of the elements in the order of their nuclear charge numbers. This conclusion had already been reached from radioactive considerations by A. van den Broek at the beginning of 1913, i.e., before any of the others, although he was wrong in assuming that the nuclear charge is always one-half of the atomic weight. The quantum theory explains the (not entirely exact) periodicity of the chemical properties; W. Kossel and especially Niels Bohr are responsible for this advance (Chapter XIV).

Modern physics believes only a few elementary particles to be indivisible. One is the electron, another the nucleus of the hydrogen atom, i.e., the proton, which was encountered first in gas discharges, for example, as canal rays. In

1932 C. D. Anderson discovered, in the Wilson cloud chamber, the positron, a particle with a positive charge and (approximately) the mass of the electron. That same year J. Chadwick found the neutron in radioactive processes. It is an uncharged particle of about the same mass as the proton. Cosmic radiation was discovered by V. F. Hess in 1910, a finding quickly confirmed by Werner Kolhörster (1887-1945). Anderson, again by means of the cloud chamber, found (1937) that cosmic radiation contains the meson, whose existence had been predicted on theoretical grounds by H. Yukawa in 1935. All indications point to its being a short-lived elementary particle, with a positive or negative charge, about 200 times as heavy as the electron, and therefore still 10 times as light as the proton. However, according to an idea put out by W. Heisenberg in 1932, and also by Ig. Tamm and D. Ivanenko, an atomic nucleus consists of protons and neutrons; its nuclear charge number states the number of its protons, and the number of neutrons is such that the sum of the masses of all the protons and neutrons produces the atomic weight. This theory, which is a product of radioactive observations and has been verified by them, has served to revive the Proutian hypothesis after more than a century. The objection that many atomic weights are not whole numbers was removed long ago.

These deviations were shown to be merely apparent by Frederick Soddys discovery in 1910 of the existence of isotopes of several atomic species. At first, he believed that this condition applied only to radioactive elements, but as time went on, it became certain that practically every position in the periodic system is occupied not by a single but several atomic species. Each of the latter has, of course, the same nuclear charge, and therefore also the same electronic arrangement in the atom and the same valence-chemical behavior, but they differ in respect to mass. The atomic weights of the individual atomic species actually are very close to whole numbers, the mass numbers, if the unit mass is taken to be not that of the hydrogen atom, but if instead the mass 16 is ascribed to the most abundant oxygen isotope. The change is not very great, inasmuch as the atomic weight of hydrogen then becomes 1.00813, but nevertheless it is of basic importance. Chemistry always deals with mixtures of isotopes and accordingly obtains only an average value that can be calculated from the true atomic weights and frequency of occurrence of the various atomic species. The slight deviations, however, that then still remain between atomic weights and their mass numbers is explained by Einsteins law of the inertia of energy (Chapter II) as the consequence of the loss of energy that is associated with the union of protons and neutrons to produce the atomic nucleus.

Isotopic atomic species cannot be differentiated by chemical methods. They can be detected in the presence of each other by magnetic and electrical deflection of canal rays, for instance, when the isotopes, because they have different masses but identical charges, travel along different paths. This mass spectroscopy began in 1898 with the deflection experiments of W. Wien. By its aid, J. J. Thomson in 1913 obtained the sensational detection of the isotopism of two atomic species of non-radioactive origin, namely the neon isotopes with the mass numbers 20 and 22 respectively. Since 1919, F. W. Aston perfected the method to the point that by 1938 no less than 260 different atomic species had been recognized (in comparison with 92 known chemical elements). In the 50th position of the periodic system, which is usually occupied by tin, there are ten atomic species with mass numbers extending from 112 to 124. (The mean atomic weight of tin determined by chemical methods is 118.8.) The first position in the system is occupied not only by the ordinary variety of hydrogen but also by deuterium (atomic weight 2.014725), discovered by H. C. Urey in 1932. Heavy water has been known since then; in its molecule, one or both of the hydrogen atoms may be replaced by deuterium atoms. In order to determine atomic weights with an accuracy equal to the one just given and thus place on a more certain basis the conclusion regarding the energy emissions during the union of protons and neutrons to form atomic nuclei, it was necessary that mass spectroscopy be developed into a precision method. This was accomplished by A. L. Dempster and particularly by J. Mattauch.

Physicists, in general, are now convinced that not only the atoms, which have a fairly complex structure, but also the elementary particles possess the full reality of other things of the external world. However, this has not been the case very long. Doubts about it persisted into the twentieth century. For example, Ludwig Boltzmann to the end of life suffered from the fact that many did not consider his kinetic theory of gases as a perfectly reliable means of explaining physical phenomena. The change that has occurred since then is due to much new knowledge. It has already been pointed out that one reason could be found in the thermodynamic fluctuations (page 115). Many other reasons could be cited, such as conferring reality on the scattered waves of each individual atom in X-ray interferences, thus arriving at the proper theory of this phenomenon (Chapter XII). But, in the main, it is certainly the Wilson cloud chamber, which gives visible evidence of the paths of individual charged elementary particles or atoms, that removed all doubts. In any case, the twentieth century brought a complete victory of atomistics.

Nevertheless, the substance concepts as regards elementary particles must be fundamentally revised. It contained the idea of uncreatableness and impossibility of annihilation; it views every portion of substance as an individual, which remains itself through all the changes it may undergo, and – if not actually in experiment, nonetheless in idea – can be identified at all times. This idea is not valid, at least with respect to the electron and positron. The researches, which followed (1933-34) the discovery of the positron, showed that if a sufficiently large γ-ray quantum strikes an atomic nucleus, an electron and a positron are formed on the nucleus simultaneously. Conversely, when electrons and positrons meet, they may mutually annihilate each other with production of γ-radiation quanta. In accord with the Einstein law of inertia of energy, the entire mass, not merely the mass at rest, but the mass augmented by virtue of the motion, is then converted to radiation energy. The newer physics has produced this result, all of whose consequences are fully recognized. It is among the most thrilling things which science has ever brought to light.

11 Nuclear Physics

Nuclear Physics

Scarcely anything has contributed so much to the change in the concept of the atom (Chapter X) as radioactivity. It was discovered by Henri Becquerel (1852-1908) in February 1896, following the published announcement just a month earlier of the discovery of X rays by Röntgen (Chapter IV).

In those days, X rays came from the fluorescent walls of the glass tubes in which they were formed, and so the idea arose that fluorescence or phosphorescence might be responsible for them. Accordingly, Becquerel tested a number of phosphorescent materials for a penetrating, photographically-active radiation. He had no success until he tried a uranium salt, but he quickly was forced to conclude that the observed radiation had no causal connection with the phosphorescence. It is now known that what he observed was the effect of fast electrons. In addition, Becquerel discovered the ionization of the air by the radiation emitted from the uranium compounds. A gigantic new field of research was thus opened. Immediately many rushed in because Röntgen's discovery had rendered the time ripe for the proper appreciation of such discoveries.

Pierre Curie (1859-1906) and his wife Marie Curie (1867-1934) were among them. This couple systematically examined all known chemical elements for radioactivity (they coined this term) and found activity also in thorium – however at the same time as Gerhard C. Schmidt – but a million times stronger in two new elements, polonium and radium. The analytical chemical Curie method, which is characterized by testing all precipitates for radioactivity, in the hands of numerous successors during the next two decades led to the discovery of the other 'natural" radioactive elements. Otto Hahn, especially, completed the list by his discovery of radiothorium (1905), of mesothorium (1906), and, together with Lise Meitner, of proactinium (1918). Somewhat different procedures were necessary only in the case of radioactive gases, the emanations. The first of these, thorium emanation, was discovered in 1900 by Rutherford.

As early as 1897, this great investigator distinguished two kinds of radioactive radiation on the basis of their penetrating powers; namely, the more readily absorbable a and the more penetrating β-rays. Whereas, the latter quickly proved themselves to be electrons by the ease with which they were deflected in an electrical or magnetic field, Rutherford had to work for years before he was able to determine the nature of the former. However, in 1903 he found from deflection measurements that the ratio of their mass and charge agreed with respect to sign and magnitude with doubly positively charged helium atoms. Thereupon, in 1904 William Ramsay (1852-1916) and Frederick Soddy proved the remarkable occurrence of helium in radium compounds, a fact which could only be explained on the basis of the production of helium from radium. In 1909 Rutherford and T. Royd confirmed the identity of a particles and helium ions by showing that collected neutralized a particles exhibit the yellow line of the helium spectrum. In this way, it was proved that the element helium can originate in other elements. About the same time it gradually became evident that, with few exceptions, a radioactive material emits either a or p rays exclusively; the non-deflectable γ radiation, discovered in 1900 by Paul Villard, may be associated with either.

However, this finding by Rutherford was not the first indication of radioactive atomic transmutation; rather this goes back to 1903. At that time considerable excitement was aroused when Pierre Curie and A. Laborde found that a specimen of a particularly pure radium salt constantly was at a higher temperature than its surroundings. The reason was a continuous production of heat, which, according to their measurement, was liberated at the rate of about 100 gram calories per gram of radium per hour. This astounding fact was substantiated later through counts of the a particles emitted per second. This number, determined in 1908 by Rutherford and Geiger, together with the energy of the individual particle as measured by its deflection, yielded on computation the same energy production that had been found by Curie and Laborde. Immediately the question was raised: What is the source of this unceasing supply of energy? As early as 1903, however, Rutherford and Soddy had conceived the idea that every radioactive process is a transmutation of elements. It was then clear that the energy given off during the individual elementary process equals the energy difference between the new and the old atom. Since then it has been customary to speak of radioactive decay, and gradually it was found proper to place all 'natural radioactive elements in three disintegration series, stemming from uranium, proactinium, and thorium respectively. Radium and polonium are in the uranium series. The insertion of these elements into the periodic system led (1911-1913) A. S. Russel, K. Fajans, and F. Soddy to the displacement laws, which state that the emission of an a particle decreases the atomic number of the atom by two, whereas the loss of a p particle raises it by one. These shifts are in complete agreement with the identity of the atomic and the nuclear charge number, a fact which was definitely established in 1913 by X-ray spectroscopy. The old view that the chemical atoms could neither be destroyed nor created was thus brought to an end.

γ radiation has no direct bearing on the transmutation of elements. It is produced only when, in the sense of the quantum theory, an excited nucleus is formed, and the latter then passes into the normal state with the emission of a γ quantum. Lise Meitner in 1926 proved experimentally that the γ radiation is not produced until after the transmutation.

One of the earliest observations in this field was that the radioactivity of a preparation diminishes with time, although the rate varies from case to case. The law of decay was found in 1899 by Julius Elster (1854-1920) and Hans Friedrich Geitel (1855-1923). It states: The number of particles emitted per second decreases exponentially with time. The constant of this law, the half-life, which characterizes each element, i.e., the time in which this number falls to one-half, varies, of course, between wide limits, from 1.6×10^{10} years for thorium to 10^{-4} second for radium C' and even smaller values. In the case of a emitters, it is connected, according to the rule discovered (1912) by Hans Geiger (1882-1945) and J. M. Nuttall, with the energy and consequently also with the range of the emitted particles. The rule states that within each disintegration series, the energy is greater the shorter the half-life period. G. Gamow gave the wave mechanics explanation of this in 1928 (Chapter XIV).

An advance of incalculable significance was made in 1905 when E. v. Schweidler furnished the explanation of the empirical disintegration law: The probability of decay is independent of time for every atom and naturally it becomes greater the shorter the period of decay. Here, for the first time, physics encountered a process that was not accessible to causality. Even now no reason can be given as to why a radioactive atom disintegrates at a particular instant and not at some other time. A veil is drawn over this mystery by the repeatedly confirmed impossibility of modifying radioactive decay through any physical means. The correctness of the probability theory was confirmed by the observations (1906-1908) made by F .W. G. Kohlrausch, Edgar Meyer, and E. Regener, and likewise by H. Geiger, of the variations, as demanded by the theory, of the number of particles emitted per unit time.

The importance of the Schweidler theory lies in the fact that subsequently many other atomic processes were discovered for which the physicist can state a probability quite well without being able to determine causally the instant of their occurrence. The Schweidler viewpoint can be carried over to all these cases.

The proof of atomic disintegration revived the alchemical notion of transmuting one element into another. Numerous attempts to accomplish this feat, by means of the electric arc, for example, were made in the decades up to around 1930. However, none of the apparent changes withstood the test of critical examination. It is now known that there is only one method of concentrating the necessary energy on a single atom: through the impact of other atomic particles or of light quanta of γ rays. But even in such experiments, the initial (1907) apparent successes were misleading. The first actual artificial atomic transmutation was accomplished in 1919 by Rutherford; he bombarded nitrogen with a particles and obtained protons of greater range. Wilson photographs of this process, such as were made in 1925 by P. M. S. Blackett, show in addition to the long track of the proton also the short track of the other product: the oxygen isotope of the mass number 17. During the period 1921-1924, Rutherford and J. Chadwick proved the occurrence of corresponding reactions, i.e., capture of an a particle and emission of a proton, with all the elements from boron (atomic number 5) to potassium (atomic number 19), with the exception of carbon and oxygen. Invariably such reactions produce, besides the proton, the succeeding element in the periodic system.

A significant year for the development of this field was 1930. First of all, W. Bothe and H. Becker observed a penetrating γ radiation when they bombarded lighter elements, especially beryllium, with a particles. This product is due, as K. Schnetzler showed in 1935, to an excitation of the beryllium nucleus through the impact, with subsequent return to the normal state with emission of one γ quantum, and is connected with a lower energy limit of the a particles of 2.3×10^6 electron volts. Irene Curie and F. Joliot verified the fundamental observation, but obtained other absorption data for the resulting radiation when they used a modified apparatus. By means of the Wilson chamber, they showed that this radiation sets lighter atoms into such rapid motion that the impulse theory cannot account for it on the basis of an effect of the γ rays. From this J. Chadwick then concluded that, in addition to the γ radiation, there is also formed a corpuscle, whose charge is zero and which has about the same mass as the proton – in other words, the neutron, whose existence had long been suspected by Rutherford. Actually, as K. Schnetzler demonstrated, there can occur, besides the excitation of the beryllium, a capture of the a particle with subsequent loss of a neutron, provided the energy of the a particle is at least 4.7×10^6 electron volts. In this case, a carbon atom will remain.

This idea proved to be entirely correct; a host of nuclear reactions were discovered in which the capture of an a particle or a deuteron led to the splitting off of a neutron. In 1934, the impact of fast deuterons against others of their own kind provided M. L. E. Oliphant, P. Harteck, and Lord Rutherford with an especially fertile source of neutrons, which made possible many neutron experiments.

In 1934 I. Curie and F. Joliot, while carrying on such experiments, happened on reactions in which the newly produced nucleus is not stable, but instead undergoes further radioactive change with the loss of a positron. Also in 1934, Fermi began his exceptionally successful experiments with neutron bombardment. The Schweidler decay law has proved its worth likewise in regard to the "artificial" radioactive atoms. Cases were subsequently encountered in which a negative electron appears in place of the positron, and under some circumstances also a γ-radiation. Since then, the number of stable or radioactive nuclei produced by artificial transformation has reached many hundreds, and almost all the places in the periodic system were filled with isotopic atomic species (Chapter X).

In all these reactions, one atom gives rise to others, which either occupy the same place in the periodic system or neighboring positions. Therefore, it was like a bombshell when in 1938 Otto Hahn and F. Strassmann showed that neutron bombardment of uranium, the last element of the periodic system, caused this element to disintegrate into elements that lie in the middle portions of this system. In fact, many kinds of fission were observed; the resulting atomic species were mostly unstable and immediately underwent further decay; the half-life period of some of them was not more than seconds, so that Hahn was compelled to develop the Curie method into a rapid procedure, which in some instances had to be completed in not more than a few seconds. It seems important that the two elements preceding uranium, namely, proactinium and thorium, also undergo a similar fission when they are subjected to the action of neutrons, although higher neutron energy is needed to initiate the effect than in the case of uranium. In addition, G. N. Flerov and K. A. Petrzhak in 1940 discovered a spontaneous splitting of the uranium nucleus with the longest half-life period known up to then, namely about 2×10^{15} years; it was detected through the neutrons liberated in the process. Hence, it was understandable why the periodic system ended with these three natural elements. Trans-uranic elements have been prepared, but they disintegrate very rapidly.

The splitting of uranium by neutrons had now made it possible also to employ atomic energy, a Jules Verne dream that had come to many, but which even Rutherford had declared to be a pure phantasy. When the uranium isotope with the mass number 235 is split by means of neutrons, on an average two neutrons are liberated at each elementary action, i.e., when one neutron is consumed. This fact was known to F. Joliot, L. Kowarski, and H. von Halban, Jr. as early as 1939. Under proper conditions, it makes possible a chain reaction, which continuously leads to more nuclear fissions, so that the process goes on of its own accord at an ever faster rate. World War II, which began eight months after the announcement of Hahns discovery, caused the first application to be the creation of a terrible new atomic weapon. A great cooperative effort of American and British scientists and engineers, supported by immense government funds, produced the atom bomb during the course of this war. Specimens were set off on July 16, 1945, in New Mexico, on August 6, 1945, over Hiroshima, and shortly thereafter over Nagasaki. In 1949, Jesse Dumond determined the wavelength of this death ray by means of a crystal spectrometer. His result, 2.43×10^{-10} cm agreed with the predicted value.

The bounds of this book do not permit a detailed account of this development. From the physical standpoint, it is the greatest experiment ever instituted by man. It was the brilliant confirmation of a bold prediction, based on a conviction of the objective truth of physics. It is not possible, at present, to hazard even a guess as to the extent to which mankind will be affected internally and externally by the consequences of this feat. Possibly, later historians will regard the dates just given as the most significant of a whole epoch. All the more so, since a retarded progress of the chain reaction, which in the bomb leads to explosion, makes it feasible to utilize for peaceful, constructive purposes, the atomic nuclear energy in the uranium piles, of which there are now several operating in America.

12 Physics of Crystals

The science of crystals belong exclusively to modern times. The regular forms of many diamonds and also the smooth faces of other crystals must undoubtedly have attracted attention from the remotest times. However, the apparent lack of regularity in the variation of their size and form most assuredly was the reason that no laws concerning crystals were discovered. In ancient times, the study of minerals, etc., did not advance beyond a purely casual state, and besides it was intertwined with mythology and superstitions about the magic power of gem stones.

A little book by Johannes Kepler, published in 1611, stands quite alone. The mere contemplation of the hexagonal snow, which gave the book its title, led this gifted man to ideas of symmetry and even to imagining that snow is built up of densely packed spheres. Trains of thought such as he employed in 1596 in his Prodromus for the derivation of a law for the radii of the planetary orbits can be discerned in his geometrical discussions. This law was quickly found not to hold, but the viewpoint that the world is the work of a spirit who rejoices in simple mathematical relationships led Kepler here along a correct path. However, this pamphlet, which was composed half in jest, made no impression even on the few who read it.

It was quite an accomplishment, when in 1669 Niels Stenson (Nikolaus Steno, 1638-1687), found that the angles between similar pairs of faces of quartz are always the same no matter how they may be developed. He found this same constancy in several other crystalline materials. The name rock crystal (for quartz) gradually came to be applied in shortened form to all solids with well-developed natural features. In this same year, Erasmus Berthelsen (Bartholinus, 1625-1698) noted that Iceland spar (calcite) exhibits double refraction of light, a fact which Huygens explained in 1678 by means of the wave theory (Chapter IV). In 1688 Domenico Guglielminis (1655-1710) extended the law of the angular constancy to several varieties of crystalline salts. However, little progress was made for a whole century. The great strides made by the rest of physics did not touch crystallography, because the physicists seldom saw well developed crystals, and the mineralogists who, of course, had plenty of such specimens were interested primarily in other problems. The difficult art of growing artificial crystals was not systematically developed until the twentieth century.

An exception is provided in the discovery of pyroelectricity in tourmaline, which, after a long period of misunderstanding, was recognized in 1758 by Franz Ulrich Theodor Äpinus as due to an electrical charge of the surfaces resulting from a temperature change.

Not until 1772 did another important work on crystal forms appear. In this, Jean Baptiste Rome de l'Isle (1736-1790) extended the law of the constancy of the plane angles to many other crystals. These angles, i.e., the position of the faces with respect to each other, are, as has been known since that time, the truly characteristic feature of each variety of crystal, whereas the size of the faces is greatly dependent on the conditions prevailing during the growth of the crystal.

With this law as a basis, geometrical crystallography was developed through tedious individual study and by no means without occasional excursions along wrong paths. However, the epoch-making studies of Christian Samuel Weiss (1780-1856), the investigations by his pupil Franz Ernst Neumann, the first physicist of note who also dealt with crystallography, and also the researches of Friedrich Mohs (1773-1889) and Karl Friedrich Naumann (1797-1873), and finally (1839) those of William Hallowes Miller (1801-1880) led to the law of rational indices." This law states that the position of each crystal face can be characterized in terms of three moderately large whole numbers, their indices," if previously three crystal axes are known together with one axial length on each. These investigators also attempted to divide crystals into systems. However, a complete systematization, the geometric demonstration on the basis of the rationality law that there are 32 crystal classes and no more, did not come until the end of that period, namely in 1830, when this advance was accomplished by Johann Friedrich Christian Hessel (1796-1872). Even his work was overlooked for decades, so that this systematization, though in a more elegant form, was set up anew in 1867 by Axel Gadolin (1828-1939), who was unaware of his predecessor. The goal of geometric crystallography was thus reached.

At first, the crystal classes were differentiated by the symmetries with respect to the positions of the boundary planes. The development of this knowledge was paralleled by a recognition that these same symmetries are decisive for the events which take place within the crystal, such as the propagation of light and elasticity. Calcite, in which double refraction was first observed, has only one optically prominent axis. Biaxiality was discovered in mica by Jean Baptiste Biot in 1812; David Brewster (1781-1868) confirmed this finding in topaz and other crystals in 1813, and in 1818 lengthened the list of double refracting materials to more than 100. The eminent astronomer, John Frederick William Herschel, supplemented this knowledge, for instance, by employing monochromatic light. But it was not until 1833 that the connection with the geometric symmetry of the position of the faces received its fundamentally correct formulation by Franz Neumann, who also referred the Fresnel crystal optics back to the elastic theory of light. This same great investigator created the theory of crystal elasticity, and his pupil, Woldemar Voigt, followed in his footsteps. The latter's *Textbook of Crystal Physics* (1910) is still the inexhaustible source of information concerning all physical questions about crystals. In it, for instance, there can be found the theory of pyroelectricity, which was given by Lord Kelvi n in 1878, and likewise that of piezoelectricity, discovered in 1881 by Pierre Curie.

Today, however, the essence of the crystalline state is not seen in any of these properties, but in the arrangement of the atoms to produce space lattices, i.e., to form configurations with strict periodicity in three directions. All the properties just mentioned can easily be understood as consequences of this idea. The space lattice theory has a long history. The packing of spheres, of which Kepler wrote in 1611, were space lattices in fact, though he did not coin this idea. The much occupied Robert Hooke (1635-1703) in his *Micrographia* (1665) stated, as is true so frequently in other fields, a correct idea of crystal structure, without giving any basis for his belief and without developing the idea. In 1690 Huygens, in his Traité de la lumire (Chapter IV) on the basis of the cleavage of calcite, assumed a space lattice composed of minute ellipsoidal particles. Because of this same characteristic, Tobern Bergman (1735-1784) in 1773, and, in a more general manner (1782 and later). René Just Haüy (1743-1822) conceived the crystal as akin to a masonry structure, constructed from the tiny parallelepipedal building stones, which likewise exhibits threefold periodicity. The first scientist, however, to combine the newly created concept of the chemical atom with this idea and to assume that space lattices are made up of chemical atoms was the physicist Ludwig August Seeber. With profound physical insight, he went even beyond this purely geometrical assumption, in that he definitely viewed the interatomic distances as being determined by the forces acting between the atoms and he related elasticity and thermal expansion to this postulate. He published his ideas in 1824, i.e., 32 years prior to the entry of atomistics into modern physics in the form of the kinetic theory of gases. But perhaps it was precisely for this reason that his feat, for it was a feat, fell into oblivion. Matters were not improved even when in 1831 the great Karl Friedrich Gauss, in discussing a mathematical book, pointed out the problems posed by Seeber's idea of the parallelepipedal arrangement of points in space. Seebers work was not disinterred until 1879 when Sohneke (see below) brought it back to light. The mathematics of the space lattice, which characterizes every such lattice, not by its content of physical structures, but purely through its congruence operations, developed entirely independent of Seeber's work.

Thus, in 1835 and 1836, Moritz Ludwig Frankenheim (1801-1869) put the question; Do the geometrically possible varieties of space lattices conform to the symmetries established in the case of crystals? Even before his second study, namely 1850, Auguste Bravais (1811-1863) derived the 14 space lattices, which bear his name, and which can be formed purely by translations of a point (without recourse to other congruence operations). These purely geometric-group theory attacks were extended in 1879 by Leonard Sohneke (1842-1897) who added certain other congruence operations and thus arrived at 65 different space groups. The complete solution of the mathematical problem, the setting up and enumeration of all the theoretically possible space groupings, is due, however, to the crystallographer Jevgraph Stepanowitsch von Federow (1853-1919) and the mathematician Artur Schoenflies (1853-1928). Independently, and by quite different methods, they both arrived in 1891 at 230 space groups.

At first, these studies had no effect on physics because no physical phenomenon required the acceptance of the space lattice hypothesis. Among the few physicists who were at all interested in crystallography, some adopted the opposite view, that in crystals, as elsewhere in matter, the molecular centers of gravity were distributed irregularly and that only the parallel placing of preferred directions in the molecules produced anisotropy. Neither was there much discussion of the hypothesis in mineralogy. Paul von Groth (1843-1927) alone upheld the Sohneke tradition in his teaching at Munich. The triumph of this hypothesis came in 1912 through the experiments of W. Friedrich and Paul Knipping who, by means of X rays, demonstrated the interference phenomena occasioned by the space lattice of the crystal, a finding which verified the prediction of M. von Laue. Because of their short wavelength, these waves are able to reveal optically the interatomic distances, whereas these elude radiations of longer wavelengths, such as light. These experiments also furnished the first decisive proof of the wave nature of X rays, which up to then had been denied by some eminent scientists because of the particularly striking quanta phenomena shown by them (Chapter XIV). The theory of this interference phenomenon, which Laue suggested in his first paper and which was verified quantitatively, is an easy generalization of the theory given by Schwerd (Chapter IV) in 1835 for optical gratings. The finding was doubted by some but not for long because the few sharp interference maxima of X rays are too suggestive of optical grating spectra. Though only an approximation, time has proved the theory to be an astoundingly close approximation. Here the wave theory of X rays and the atomic theory of crystals come together, one of those surprising events to which physics owes its powers of conviction.

This theory permits a comparison of the wavelength with the three period lengths of the space lattice. Since at first it was possible to state no more than their order of magnitude, an absolute determination of the wavelength was impossible. The difficulty lay in the ignorance of the atomic structure; it was not known how many atoms resided in the individual space lattice. At this point, in 1913 William Henry Bragg (1862-1942) and his son William Lawrence Bragg brought aid in the form of an hypothesis that had been set up in 1898 by William Barlow (1845-1934) concerning the structure of rock salt, NaCl. Once again, the idea of the densest spherical packing plays a part here. The Braggs confirmed this structure by means of the intensities of the interference maxima, and thus obtained an absolute measure for the lattice constant, and thereupon they could determine the wavelengths of the X rays in absolute terms, i.e., in centimeters. With this aid, they were also enabled to make absolute determinations of the lattice constants of other crystals. Usually these lie between 10^{-8} and $10-7$ centimeter; however, considerably larger values are found for complicated organic compounds. In 1923 A. H. Compton observed the diffraction of X rays by artificial gratings, but his measurements added nothing to the earlier determinations of wavelengths except a very considerable increase in accuracy.

Measurement of the wavelength created X-ray spectroscopy. In 1908 C. G. Barkla and C. A. Sadler distinguished, by means of their different absorbabilities, the characteristic K, L, M $\cdots$ radiations of the chemical elements. Since 1913, initially through studies by the Braggs and H. G. J. Moseley (1887-1915), these radiations have been resolved into series of sharp spectral lines, whose wavelengths, independent of the chemical binding, exhibit simple relationships to the position in the periodic system. These radiations attained great importance with regard to atomistics (Chapters X and XIV) and besides led to the discovery (1923) of hafnium by G. von Hevesy and of rhenium (1925) by W. Noddack, J. Tacke, and Otto Berg (1874-1939.) The study of crystal structures, in which L. A. Seebers idea finds its brilliant confirmation, has become a distinctly important branch of physics. The number of organic and inorganic crystals for which the atomic positions can now be accurately stated is in the thousands. Included are such complicated structures as those of the various silicates, the earliest being garnet, which was studied in 1925 by G. Menzer. Many metals, such as aluminum, silver, and copper, have been found to conform to the densest spherical packing, a verification of the thought published in 1611 by Kepler. A röntgenographic Fourier analysis of the electron density, proceeding from the establishment of the atomic centers, may serve also to give considerable information about the electron distribution in these and other not too complicated structures. This was known to W. H. Bragg in 1915.

X-rays have now also revealed the wide distribution of the crystalline condition. Of course, well-developed large crystals are relatively scarce; a microcrystalline structure of microscopic or still smaller crystallites in thorough confusion is met far oftener. This idea of crystallinity is an old one with respect to metals. However, the fact that wood, textile, muscle. and nerve fibers are likewise microcrystalline is new. In truth, the crystalline state is the normal condition of solids; only a few substances, the glasses in particular, are exceptions to this rule. Hence the entire atomic theory of solids, for instance the quantum theory of electrical conduction, emanates from the space lattice.

Space lattices performed a special mission after the establishment of wave mechanics (Chapter IV). In 1925 W. Elsasser deduced from L. de Broglies theory that beams of electrons passing through crystals must show interference phenomena just as X-rays do. This expectation was verified in 1927 by experiments carried out by C. J. Davidson and L. H. Germer on one hand and by G. P. Thomson on the other. They thus furnished directly evident proof for this revolutionary theory and at the same time they strengthened it quantitatively by measuring the wavelengths of these rays. Corresponding results were obtained for helium and hydrogen atomic radiations of lesser energy (hundredths of electron volts) by Otto Stern (1929) and Th. Johnson (1931), likewise with the aid of crystals. In these cases, however, only the surfaces are active, since these rays do not penetrate solids. In contrast, space lattice effects have been revealed with neutrons just as plainly as with X-rays ever since American scientists have had access to the powerful sources of neutrons provided by the uranium piles (Chapter XI).

It should be pointed out that X-rays and electron interferences can be applied also to the determination of the form and size of simple gas molecules. This was demonstrated by P. Debye in 1915 with X-rays and by H. Mark and R. Wierl (1903-1932) with beams of electrons. These researches have furnished the distances between the atomic nuclei for many diatomic molecules, such as nitrogen, oxygen, and fluorine; they lie between 1×10^{-8} and 3×10^{-8} centimeter. It is known that the molecule of carbon dioxide is linear, whereas that of water is angular and so on. The molecule of carbon tetrachloride, CCI_4 , has been measured especially well; the chlorine atoms form an equilateral tetrahedron, whose center is occupied by the carbon atom. The stereochemical concepts announced by J. H. vant Hoff in 1874 are thus completely substantiated.

The original theory of space lattice interferences presents, as was stated above, only an approximation which suffices in practically all instances for X-rays, but frequently fails for electrons. Its extension into a more exact dynamic theory for X-rays was furnished in various forms in 1914 by C. G. Darwin and in 1917 by P. P. Ewald, who, with its aid, were able to explain the deviations of the precision measurements made by W. Stenström (1919) from the earlier theory. The dynamic theory received its final form in 1931 at the hands of M. von Laue; its union with wave mechanics was accomplished in 1935 by M. Kohler. The corresponding advance for electrons had been made as early as 1928 by H. Bethe.

The dynamic theory, in contrast to its predecessor, describes also the waves in the interior of the crystal. Consequently, it is essential to the understanding of the interference effects, discovered in 1935 by W. Kossel, in the emission of monochromatic X-rays by crystals, in which the radiation sources accordingly reside in the space lattice itself. This radiation exhibits characteristic sharp maxima or minima of intensity in directions which are determined by interference conditions.

The original theory was likewise incomplete in so far as it completely disregarded the thermal motion of the atoms, even though the latter, in comparison with the three periods of the space lattice, are certainly not inconsiderable at room temperature and above. P. Debye in 1914 demonstrated that this factor had no influence on the position and sharpness of the interference maxima, but it does decrease their intensity. His theory has subsequently been tested by others. It was verified by W. L. Bragg and his associates through extensive series of measurements (1926-1933).

13 HEAT RADIATION

Heat radiation is one of the youngest branches of physics. The concept was established by the chemist Karl Wilhelm Scheele (1742-1786). The first experiments along this line were carried out by Marcus Auguste Pictet (1752-1825) and from these Pierre Prévost (1751-1839) in 1791 drew the conclusion that every body emits radiation independent of its surroundings: The quantity of heat supplied to it by radiation is equal to the difference between that which it receives from the surroundings and what it emits. This is a truly remarkable law, to which there is nothing that corresponds in the conduction of heat. During the first half of the nineteenth century, the spectrum (heat and light radiation) was recognized to be a unit (Chapter IV), and since at about the same time the two first fundamental laws became known, thermodynamics and optics were thus matured to the point that their union could produce a child destined to bring about the greatest revolution of physics. There occurred here one of those events, alluded to in the Introduction, which prove the objective truth of physics.

It was an epoch-making discovery when Gustav Robert Kirchhoff found (1859) that in every cavity surrounded by walls at the same temperature, there is established a universal radiation which depends solely on the temperature and not on the nature of the material that constitutes the lining of this cavity. Furthermore, he found that the intensity of radiation of every body can be referred back to this cavity radiation if the absorption and refractive index are known.

The cosine law of directional distribution, deduced in 1760 by Johann Heinrich Lambert (1728-1777) from observations of the radiation from light sources, holds strictly only for this cavity radiation. Hence the entire problem of radiation was reduced to the investigation of cavity radiation. At that time, no one grasped the import of this law; direct study of the cavity radiation seemed to be impossible until Otto Lummer (1860-1925) and Wilhelm Wien, in 1895 discovered the device of providing the cavity with a small opening which would not affect the radiation state intrinsically. Only since then has it been possible to make quantitative measurements of the intensity of the cavity radiation.

A greater impression on their contemporaries was made by a related discovery announced a few months previously by Kirchhoff and Robert Bunsen (1811-1899), namely that the dark Fraunhofer lines in the solar spectrum coincide with the emission lines of well-known gases and vapors. This discovery showed in comprehensive measure that matter in extraterrestrial space is made up of the same chemical elements as are present on the earth, a fact that hitherto had only been indicated by studies of the composition of meteorites. Spectroscopy as an astronomical instrument immediately promised a never-dreamed-of extension of knowledge of the fixed stars; the successes, however, soon exceeded all expectations. Through a reversal of the usual order, the element helium was discovered in the sun by Jules Jansen (1824-1907) in 1868, whereas it was not until 1895 that it was found enclosed in the mineral cleveite by William Ramsay (1852-1916) and Per Teodor Cleve (1840-1905). The spectroscopy of the heavenly bodies is by no means complete at present.

Kirchhoff thought that the relation between this discovery and thermodynamics to be much closer than it really is. He erred in assuming that the emission of the spectral lines proceeds at the expense of heat energy. In the majority of cases, electrical or chemical excitation produces luminosity in gases; the temperature of the radiation, as it now is expressed, then lies far beyond that of the gas itself. The coincidence of

absorption and emission lines eventually is based on a resonance phenomenon whose final formulation was possible only
by means of the quantum theory.

The second advance in the investigation of thermal radiation was due to Ludwig Eduard Boltzmann in 1884. From
the electromagnetic theory of light he drew the conclusion
that the cavity radiation exerts against the walls of the cavity a pressure that is equal to one-third of its energy per
unit volume. By means of simple application of ordinary
thermodynamic reasoning, he thus was able to deduce that
this energy is proportional, with a universal factor, to the
fourth power of the absolute temperature. This provided a
basis for and defined a result derived by Josef Stefan (1835-
1893) from measurements made by French physicists in 1879,
and it represented, besides, a triumph of the electromagnetic
theory of light. In his eulogy of Boltzmann, H. A. Lorentz
declared that this short paper, whose well thought out dar-
ing lies in the translation of the thermodynamic concepts of
pressure and temperature (by this he also implied entropy)
to the cavity radiation, is one of the treasures of theoretical
physics.[1]

The Stefan-Boltzmann law deals with the total energy of
the whole spectrum. The determination of the energy distribution in it had to be made the objective of research. A
fundamental approximation to this constituted the third advance in the theory of heat radiation. This was accomplished
in 1893 by Wilhelm Wien by applying a combination of thermodynamic considerations with the Doppler principle to the
compression of the radiation contained in a cavity. The Wien
displacement law, a brilliant achievement, which has received
too little appreciation in the textbooks of today, makes it
possible to calculate the energy distribution for every temperature, as soon as it is known for one. Even without this
knowledge, the law suffices to explain why the intensity maximum in the spectrum shifts more and more to the shorter

---

[1]H. A. Lorentz, *Verh. d. deutschen physikalischen Gesellschaft*
(1907).

wavelengths with rising temperature, and therefore why heat radiation remains invisible at low temperatures, but has its maximum in the visible at about 6000, and makes it possible, as soon as its position is known, to determine the temperature of the radiator, the sun, for instance. Likewise, Wien was the first to extend the concept of entropy, not merely to cavity radiation but also to directed radiation, which, because of the law of the increase in entropy, was absolutely necessary, especially since the entropy of a radiator decreases. As soon became apparent, the displacement law went as far as classical physics was able to advance, i.e., to the threshold of the quantum theory.

Many attempts had been made to solve the problem of calculating the intensity as a function of the frequency and the temperature. Among these is the law, named after Lord Rayleigh (1842-1919) and James Hopwood Jeans (1877-1946), which states that the intensity is proportional to the temperature and to the square of the frequency. This cannot possibly hold for frequencies of unlimited values (short wavelengths) because then it yields no finite total amount of energy of the radiation; but the law does contain a certain measure of truth in so far as it applies to low frequencies (long wavelengths). Hence, since 1896, W. Wien, and later also Max Planck, substituted a distribution law, according to which the intensity is supposed to fade away exponentially as the wavelength decreases. This not only avoided the ultraviolet catastrophe,' but actually seemed to be well supported by experiment. However, in 1899 Otto Lummer and Ernst Pringsheim found definite departures which led Planck to renewed consideration of the subject.

In this Planck had the advantage of twenty years of activity in thermodynamics and a clear appreciation of the significance of entropy, which at that time was still greatly misunderstood. He felt that the core of the problem was not the intensity formula itself, but the definitely associated relation between the energy, the frequency, and the entropy of the radiation. One such relationship corresponded to the

Wien distribution law, another to the Rayleigh-Jeans law. In October, 1900, Planck learned that Ferdinand Kurlbaum (1857-1927) and Heinrich Rubens (1865-1922) had made new measurements which verified the latter law for long waves. He thereupon set up an interpolation formula between the two relationships, from which there resulted directly the radiation law that bears his name, and which contains the earlier formulas as limiting cases.[2] He reported this to the Deutsche physikalische Gesellschaft on October 19, 1900. Despite some opposition, the result has subsequently been verified by experiment to an increasing extent.

The main question still remained, namely, the problem of an appropriate theoretical establishment of this law that had been discovered semi-empirically. Planck went back to the relationship, revealed by Boltzmann, between entropy and probability (Chapter X) and computed the latter for an oscillator with the frequency ν. He did this by employing the unheard-of new idea, inspired solely because of necessity, that only discrete energy stages are possible. This mode of attack actually produced a radiation law. However, if this was to suffice for the Wien displacement law, each of the energy stages would have to exceed the others by an amount $h\nu$, where h is a new universal constant, the elementary quantum of action. In this way, the theoretical radiation formula became identical with the one found by interpolation. Comparison with the measurements gave the numerical value of h as 6.5×10^{-27} erg/sec, that of the Boltzmann constant, which since the Boltzmann entropy-probability relation was employed also is included in the radiation law, as 1.37×10^{-6} erg/sec. This derivation was reported to the Deutsche Physikalische Gesellschaft on December 14, 1900. The quantum theory dates from that day.

Planck's attachment of $h\nu$ to the concept of energy did not merely constitute another extension to the existing physics, it was a real revolution. The succeeding decades have re-

---

[2]See in this connection M. Planck, Zur Geschichtc der Auffindung des physikalischen Wirkungsquantums. *Naturwissenschaften* 31, 153 (1943).

vealed with increasing clarity not only the depths of its import, but also how greatly it was needed. The assistance of the quantum idea has made it possible to arrive at an understanding of all atomic processes which hitherto had remained closed to physics.

Time has brought other theoretical deductions of the Planck radiation law. For instance, in 1910 P. Debye applied the $h\nu$ factor for energy to the electromagnetic characteristic vibration of a cavity and in this way arrived, perhaps more simply, at the radiation formula. The writer considers the Einstein derivation (1917) as still more remarkable. It is most widely separated from the vibration concept of cavity radiation. It characterizes the latter by means of spectral regions and the energy quanta, which belong to these regions. It also assigns to every excited atom in the cavity a definite probability of radiating per unit time, but likewise also gives it a probability, proportional to the energy of the radiation, of being absorbed or of being forced to emit. It attaches only an absorption probability to non-excited atoms. The Schweidler postulate of the disintegration probability in radioactivity is here given its adaptation to other atomic processes; it has been extended over the entire quantum theory.

The thermodynamics of radiation yielded as a by-product a surprising confirmation of the Boltzmann principle. Two partial systems that are separated in space are statistically independent in general so that their probabilities are multiplied when that of the whole system is computed. To this there corresponds, according to this principle, the additive union of their entropies to obtain the total entropy, which belongs to the common tacit assumptions of classical thermodynamics. If this computation is made in the case of the two coherent rays that are obtained from one ray by reflection and refraction, their total entropy will be found to be greater than that of the original ray. However, M. von Laue, in 1906, was able to show that this process is reversible, i.e., two coherent rays, by appropriate reflection or refraction, can be reunited into a single ray. The total entropy of the two co-

148

herent rays must logically be equal to that of the original ray. The contradiction can be resolved if the additivity of entropy is abandoned. According to the Boltzmann principle this is actually necessary because one of the two rays, with respect to all the details of its oscillation, is determined by the other, it is not statistically independent of it. This single exception to the additivity of entropies would be unintelligible without the Boltzmann principle.

14 Quantum Physics

Quantum physics, which in contrast to the earlier theory, is characterized by the appearance of the elementary action quantum h and the designation of states in material systems by *whole* numbers, dates as theory only from the beginning of the twentieth century (Chapter XIII). However, some of its experimental roots extend far back into the nineteenth century. Of course, the measurements of the intensity of heat radiation which brought the change are a product of the last decade of that century. However, the photoelectric effect, and the wave lengths of the line and band spectra and also the dependence of the specific heats of certain substances on temperature had been known decades earlier. The older physics had hoped to arrive at an explanation of these findings; otherwise it is difficult to understand why Philipp von Jolly (1809-1884) told the inquiring young Planck that physics was essentially worked out and the pursuit of this science accordingly could hardly be very profitable. What appeared from time to time concerning line spectra could no longer stand up under rigid criticism when the discussion was based on the older ideas. On the other hand, quantum physics handled these problems more or less easily and in addition elucidated much of the newly acquired experimental observations.

At first Planck's radiation theory received little attention. The idea of discrete energy changes was entirely too novel. Some may also have become suspicious because his determination of the electrical elementary quantum from radiation measurements (Chapter X) was considerably higher than the majority of the current values which were being obtained by direct methods. The first bold step toward the further promotion of the quantum idea was taken by A. Einstein in 1905, when he brought out his theory of the photoelectric effect.

The earliest indication of this effect was noted in 1887 by Heinrich Hertz. He found that the discharge is facilitated when ultraviolet light falls on a spark gap. One year later, Wilhelm Hallwachs (1859-1922) proved that the reason for this was the presence of carriers of electricity, whose nature as free electrons was clarified in 1899 by Philipp Lenard. In 1902 the latter announced two astonishing rules concerning this effect. They state: The energy of the electrons rises from a lower limiting value as the frequency of the disengaging light increases and is independent of the intensity of the light, which merely determines the number of electrons set free per unit time.

These facts previously could not be explained by the undulatory theory of light, but now conformed exactly to expectation, if, according to Einstein, light is regarded as a stream of light quanta (photons), each quantum being given energy $h\nu$, and if it is further assumed that each electron is liberated by one quantum. With such assumptions, the phenomenon is the direct evidence of the bombardment of the irradiated body by the quanta. If $h\nu$ is less than the work required to liberate an electron, the effect cannot occur; hence it actually has a long wave limit in the spectrum, which still depends on the irradiated body. If, however, $h\nu$ is greater, then the energy of the photoelectron is equal to $h\nu$ diminished by the work required for the release. Einsteins theory conforms so exactly to the phenomenon that in 1916 R. A. Millikan was able to make an accurate determination of the value of h from measurements of the frequency and the electron energy.

The same ideas were used by Einstein in 1912 when he set up the fundamental law of photochemistry. According to this, every photochemical reaction consists primarily in the absorption of one quantum of light and the transformation that is thus initiated on one atom or molecule. This law likewise finally proved to be correct, after endless patience and acumen contributed by many workers, especially Emil Warburg (1846-1931) and James Franck, had elucidated all the subsequent reactions and other complications that often occur in conjunction with the elementary action just described and that makes the number of transformed molecules appear to be sometimes less and sometimes many thousand times greater than stated by the law.

The reverse of the photoelectric effect is the production of radiation through retardation of an approaching electron at an atom or molecule. If it results in one elementary action, a light quantum is produced, whose energy $h\nu$ is equal to the kinetic energy of the electron (increased by an amount corresponding to the work of release). When X rays are formed in an X ray tube, the retardation of the electrons at the anticathode in general results in several elementary actions. However, in any case, the highest possible frequency or the shortest possible wavelength corresponds to the electron energy. This is stated in the law, discovered in 1915 by W. Duane and F. L. Hunt, of the short wave limit of the Brems spectrum. Since this was not known in 1912 when X ray interferences were discovered, M. von Laue had to expect far more interference points according to his theory than actually appeared, and he erroneously ascribed the result to selective properties of the crystal-atoms. According to the Duane-Hunt law, the short wavelengths, which von Laue expected to appear at the missing points, would really not be present.

The reality of the light quantum is revealed perhaps still more plainly in the Compton scattering of X rays, discovered in 1923 by A. H. Compton, in so far as not only the energy of the photon but also impulse plays a part here. Even Röntgen had noted that these rays undergo diffuse scattering in all materials, and the fact that this scattering occurs, in part, with no change in wavelengths, as had long been known in the case of light, was one of the fundamental suppositions that led to the success of the interference experiments with crystals. But Compton showed that a scattering with increased wavelength, i.e., diminished frequency, also occurred. His theory for this, which was also discovered independently by P. Debye, is nothing else than the application of the law of conservation of energy and impulse to reciprocal action between a light quantum and a free electron. The quantum carries with it a certain energy and a certain impulse. After the collision, a part of the energy and the impulse resides on the electron and the quantum flies away with diminished energy, i.e., lowered frequency and changed direction. This idea has been found to meet the demands of all experiments along this line.

However, the preceding recital has gotten ahead of the historical development and it is necessary to pick up the thread at an earlier date.

In 1875 Heinrich Friedrich Weber (1842-1913) found that the molar heats of the two varieties of carbon, namely diamond and graphite, and also those of boron and silicon are much lower than corresponds to the Dulong-Petit law (Chapter X). It was also found that the values approach the expected value more and more as the temperature is increased. In 1907 A. Einstein, who had attended Webers lectures while a student at Zurich, provided the theory for these findings. According to Boltzmann-Gibbs statistics, the energy of harmonic resonators is a linear function of the absolute temperature, and hence the specific heats of a system consisting of such resonators is constant. According to the Planckian statistics of the resonator, however, the energy decreases with falling temperature much faster; the specific heat decreases exponentially until it reaches zero. Since Einstein ascribed to atoms of the solid, fixed positions of rest, around which each oscillates with definite frequency, he was thus able to explain qualitatively the observed decrease. The missing parts of this concept were supplied in 1911 by P. Debye, when in applying his theory of the radiation law (Chapter XIII) to the elastic characteristic vibrations of the solid, he ascribed to each the energy designated by Planck for the resonator. Thus when the absolute null point is neared, there results the famous law of the proportionality of the molar heat to the third power of the temperature. Measurements by W. Nernst and others subsequently confirmed this law for many materials.

The year 1913 brought three advances. First, J. Franck and G. Hertz discovered the stepwise retardation of electrons by gaseous atoms; the energy transfer from the colliding electron to the struck atom proceeds only in definite discrete amounts that are dependent on the atom. The explanation was ready to hand: The atoms possess discrete energy states, exactly as Planck had postulated for the resonator, except that the energy levels are not equidistant. If an atom, while at the lowest level, i.e., at the ground state, should be excited at all, the electron would need to be capable of transmitting to it the energy difference until a higher energy level is reached, and the electron then loses exactly this amount of energy. These same investigators also showed that the energy taken from the electron is often emitted as light quanta, and that the frequency of this radiation ν can be calculated from the equality of the quantum energy $h\nu$ and the transferred energy. Here then, the hitherto hypothetical discrete energy levels received direct empirical verification.

The second great experimental discovery of 1913 was due to Johannes Stark, who showed that the spectral lines of hydrogen can be resolved in an electric field. More important than the other two was the theoretical discovery of the atomic model by N. Bohr. It was a modification of the Rutherford model in that it included quantum conditions. Whereas the Rutherford model accepted a continuous succession of orbits for the movement of an electron around the nucleus of the atom, these conditions sorted out a discrete succession of circular orbits; according to the generalization by A. Sommerfeld, ellipses were also permitted. The quantum conditions specified were: the phase integrals for every permissible orbit are whole multiples of the action quantum h. Since the energy of motion is also fixed with every orbit, the result was a theory of discrete energy levels. If now the atom goes from a higher level E_1 to a lower level E_2 with emission of a quantum, this emission, in conformity with the ideas verified in the photoelectric effect, must, of necessity, have the frequency: $\nu = \frac{E_1 - E_2}{h}$ whereas, conversely, by absorption of a quantum with the energy $h\nu$, the atom goes from E_2 to E_1 This idea was proposed as early as 1912 by J. J. Thomson for the characteristic K, L, M radiation of the elements. According to Bohr, it is in this way that the line spectra are generated.

The first triumph of this theory was Bohrs interpretation of the hydrogen spectrum. In 1885 Johann Jakob Balmer (1825-1898) stated that the frequencies of the lines in the visible region are proportional to $1/2^2 - 1/m^2$, in which m is given the values $3, 4, 5$, etc., in succession. Bohr now found for his circular orbits, and likewise Sommerfeld for the permitted elliptical orbits, the discrete energy levels to be proportional, with a universal constant, to $1/m^2$ so that the frequency, according to the above formula, exactly satisfies the Balmer formula. Furthermore, the proportionality factor, the Rydberg constant, was in agreement with the measurements made by F. Paschen (Chapter IV). The Sommerfeld version of the original theory proved superior in that it permitted several orbits for each energy level of the undisturbed atom. When disturbed by an electrical or magnetic field, the various orbits of the originally uniform level receive somewhat different energies, the level splits, and, according to the formula cited above, a resolution of the spectral lines corresponds to this. Thus the theory of the Stark effect was made possible, and it was announced in 1916 by Karl Schwarzschild (1873-1916) and P. S. Epstein. The same is true of the theory of the normal Zeeman effect, which likewise was published in 1916 by P. Debye and A. Sommerfeld. If more than one electron revolves around the nucleus of the atom, as is the case for all elements with the exception of hydrogen, ionized helium, and other polyionized atoms, the calculation of the quanta orbits and the energy levels is only approximate. But even then, the

Bohr model of the atom furnishes a general systematic guide to the line spectra, including those in the X ray region. In the same way, the quanta conditions make possible a systematization of the band spectra emitted by polyatomic molecules. In the light of this theory, the experimental data gathered by spectroscopists through decades of studies afford deep insights into the arrangement of the electrons surrounding the nuclei of the atoms.

This also initiated, after preliminary work by W. Kossel in 1916, an understanding of the very mysterious periodic system of the elements (Chapter X). In 1913 X-ray spectroscopy had definitely proved that this system presents an arrangement according to nuclear charges. But how were the approximate periodicity of chemical properties and line spectra to be explained? The secret of this question was fully unveiled in 1925. S. Goudsmit and G. E. Uhlenbeck, on the basis of spectral experiments, ascribed to the electron a magnetic moment and a rotational impulse of a certain magnitude that was closely related to the Planck constant, and that same year W. Pauli set up the exclusion principle that no two electrons of a given atom can be identical with respect to all quantum numbers. A simple chain of thought, checked at each step by spectral observations, then showed why the first periods of the system each contains eight elements, the next eighteen, a later one thirty-two, and why also each period commences with an alkali metal and ends with a noble gas. Once again, two entirely different sets of ideas – the old chemical and the quantum theoretical – had unexpectedly converged and readily fused together.

The theory of magnetism likewise received an entirely new impetus through the Bohr model of the atom; the revolution of the electrons along definite orbits revived Amperes hypothesis of molecular currents (Chapter V). There was now added a statement concerning the magnitude of the moment of each elementary magnet; it is an integral multiple of the Bohr magneton, which again is closely allied to the Planck h. The correctness of this theoretical conclusion was confirmed in 1921 by W. Gerlach and O. Stern, by means of the magnetic deflection of rays of silver atoms, in which the moment of this atom proved to be exactly equal to one magneton.

Despite its great and lasting successes, the Bohr theory contained, nevertheless, a systematic defect. It employed classical mechanics for determining electron orbits, but thereupon, without inner connection with this calculation, discarded, with the aid of the quanta conditions, the overwhelming majority of these orbits as not being realized. Wave or quantum mechanics, founded in 1924-1926, is more uniform and somewhat more successful in accounting for spectra. It has now completely replaced its predecessor.

The first step was taken by Prince Louis de Broglie in 1924. On the basis of relativistic considerations, he coordinated every movement of a mass point with a wave, whose wavelength can be computed from the mechanical impulse of the particle by means of the Planckian h. Entirely different considerations led E. Schrödinger, in 1926, to set up a partial differential equation, similar to the wave equation, for such a coordinated wave, and he proved that a discrete number of energy states can be deduced from it and suitable limiting conditions. Here again, the same energy level was obtained for the hydrogen atom as from the Bohr theory, so that the Balmer formula for the hydrogen spectrum can be derived as well as from the Bohr theory. Meanwhile, in 1925 M. Born, W. Heisenberg, and P. Jordan had created a quantum mechanics, which though it appeared entirely different at first, nevertheless was mathematically identical with the Schrödinger theory, a fact pointed out by Schrödinger almost at once, in 1926. This theory also included the de Broglie relation between wavelength and impulse.

The writer believes that the physical content of this theory is not yet' fully comprehended, but it nonetheless has been applied mathematically in masterly fashion. It is supported, first of all, by all spectroscopic data, a fact of special significance since the accuracy of such determinations ranks unusually high among physics data, exceeding even the famed astronomical measurements in this respect. However, there is a more evident, though less exact proof of material waves. As was surmised in 1925 by W. Elsasser (Chapter XII), when electron rays fall on crystals, they produce interference phenomena similar to those given by X rays. This was confirmed in 1927 by C. J. Davisson and L. H. Germer, and likewise by G. P. Thomson. Otto Stern in 1929 and Th. H. Johnson in 1931 had like success in diffraction experiments with rays of helium atoms, hydrogen atoms or molecules. In fact, all these experiments verified the de Broglie formula quantitatively.

The accomplishments of this theory quickly accumulated. A particularly striking success was its application to radioactive disintegration in the case of α-rays. According to this theory, there should be a tunnel" effect, namely, the crossing of a threshold potential by a particle, whose energy according to classical mechanics is not sufficient to carry it across this barrier. In 1928 G. Gamow ascribed the emission of a particles to this tunnel effect. His theory states that a threshold potential surrounds the atomic nucleus, but the a particles possess a certain probability of passing beneath" it. This explained the relation between range and half-life found empirically by Geiger and Nuttall (Chapter XI).

Finally, it was of historical importance when wave mechanics ascribed chemical bonding between like or at least similar atoms to exchange energy." This idea was first advanced in 1927 by W. Heitler and F. London in a classical study of the hydrogen molecule H_2. This energy, which has no analogue in the older physics, is a necessary mathematical consequence of the Schrödinger wave equation. By means of the same concept, applied to the conducting electrons in metals, Werner Heisenberg in 1928 solved the age-old riddle of ferromagnetism. It is an exchange phenomenon," which arises because the magnetic moments of these electrons are ranged in parallel in iron, nickel, cobalt, but not in the other metals.

The further development of the quantum theory, which concerns the question, for instance, of the compatibility of the wave and the corpuscular ideas, is too recent to be treated historically. The limits imposed by the author for the composition of this book have been reached. In any case, one of the characteristic features of the present-day quantum physics is that it supplies no process with anything other than the probability that it will occur in a certain period of time. It computes, for example, the probability of the release of an electron by light of definite intensity and vibration frequency. A causal determined assertion lies beyond its possibilities. It thus contains a feature, which was first discovered by E. von Schweidler (Chapter XI) in the case of radioactive transformations, and was subsequently carried over by A. Einstein to the absorption and emission of light. However, the laws of the conservation of energy and impulse retain their rigid validity in quantum physics.

After 1900 Planck strove for many years to bridge, if not to close, the gap between the older and quantum physics. The effort failed, but it had value in that it provided the most convincing proof that the two could not be joined. The consequence of this is Niels Bohr's theory of the complementarity of the older corpuscular concept of the elementary particles and the quantum mechanics representation as waves (1927). Many of the physicists of today have accepted this doctrine. What position the future will take is no concern of a history of physics.

Authors

Äpinus, 52, 134

Abbe, 42

Abrahams, 68

Adelsberger, 13

Aepinus, 52

Airy, 80

Ampère, 58, 61

Anderson, 120

Arago, 45, 58

Archimedes, 2

Aristarchus, 73

Aristotle, 1, 17, 35, 71, 72

Armati, 41

Arrhenius, 4, 57, 116

Aston, 121

Avogadro, 111

Bürgi, 5

Babinet, 82

Bacon, 41

Balmer, 156

Barkla, 139

Barklas, 47

Barlow, 138

Bartholinus, 134

Becker, 129

Becquerel, 125

Bell, 29

Berg, 139

Bergman, 136

Bernoulli, 22, 97

Berthelsen, 134

Bethe, 105, 141

Bezold, 64

Biot, 39, 44, 58, 96, 135

Black, 93, 94

Blackett, 129

Bohr, 43, 119, 155, 160

Boltzmann, 64, 113,

114, 122, 145, 147

Born, 158

Bothe, 129

Boyle, 4, 7, 44

Bradley, 43

Bragg, 138, 139, 141

Braggs, 139

Brahe, 34

Bravais, 137

Brewster, 135

Briggs, 5

Brown, 115

Brugmans, 53

Bruno, 3, 35, 72

Bucherer, 25

Bunsen, 144

Carnot, 95, 98, 107

Cauchy, 22, 63

Cavendish, 4,

INDEX